THE UTILIZATION OF SECONDARY AND TRACE ELEMENTS
IN AGRICULTURE

Developments in Plant and Soil Sciences

The Utilization of Secondary and Trace Elements in Agriculture

Proceedings of a Symposium organized jointly by the United Nations
Economic Commission for Europe and the Food and Agriculture
Organization of the United Nations at Geneva, 12 – 16 January, 1987

UNITED NATIONS: Economic Commission for Europe
and Food and Agriculture Organization

1987 **MARTINUS NIJHOFF PUBLISHERS**
a member of the KLUWER ACADEMIC PUBLISHERS GROUP
DORDRECHT / BOSTON / LANCASTER
for
THE UNITED NATIONS

Distributors

for the United States and Canada: Kluwer Academic Publishers, P.O. Box 358, Accord Station, Hingham, MA 02018-0358, USA
for the UK and Ireland: Kluwer Academic Publishers, MTP Press Limited, Falcon House, Queen Square, Lancaster LA1 1RN, UK
for all other countries: Kluwer Academic Publishers Group, Distribution Center, P.O. Box 322, 3300 AH Dordrecht, The Netherlands

ISBN 90-247-3546-7 (this volume)
ISBN 90-247-2405-8 (series)

V

Contents

VI

REPORT OF THE SYMPOSIUM ON THE UTILIZATION
OF SECONDARY AND TRACE ELEMENTS
IN AGRICULTURE

Introduction

1.   The symposium, held in the Palais des Nations, Geneva, from 12
to 16 January 1987, was attended by 26 experts representing
governments, research institutes and industry of the following
countries: Belgium; Denmark; Finland; German Democratic
Republic; Germany, Federal Republic of; Hungary; Netherlands;
Norway; Poland; Portugal; Romania; Sweden; Switzerland;
Turkey; United Kingdom; the Union of Soviet Socialist Republics
and Yugoslavia. Morocco participated under Article 11 of the Terms
of Reference of the UN Economic Commission for Europe.

2.   Representatives of the following non-governmental organizations
also participated in the Symposium: International Centre of
Chemical Fertilizers (CIEC); International Potash Institute.

3.   The meeting was opened by the Director of the FAO/ECE
Agriculture and Timber Division.

Adoption of the Agenda

4.   The provisional agenda (document AGRI/SEM.21/1/Add.1) was
adopted.

Election of Officers

5.   Professor A. Finck (Federal Republic of Germany) was elected
Chairman and Dr. H. Gembarzewski (Poland) was elected Vice-Chairman
of the Symposium.

The Programme

6.   The programme of the Symposium was composed as follows:

I.   <u>Utilization of secondary nutrients in agriculture</u>
      (calcium, magnesium, sulphur, etc.)

      Rapporteurs: Mr. L.-G. Nilsson, Swedish University of
      Agricultural Sciences, Uppsala/Sweden; Mr. D. Hewgill,
      Ministry of Agriculture, Fisheries and Food, Newcastle
      upon Tyne/United Kingdom; Dr. I.A. Shilnikov, Head of
      Laboratory at the All-Union Research Institute for
      Fertilizers and Soil Science, Moscow/USSR and Dr. H.
      Gembarzewski, Fertilizer and Soil Research Institute,
      Wroclaw/Poland

II.   <u>Utilizaton of trace elements (micronutrients) in
      agriculture</u>
      (boron, iron, zinc, copper, manganese, molybdenum, etc.)

      Rapporteurs: Professor Dr. M. Sillanpää, Director,
      Institute of Soil Science, Agricultural Research Centre,
      Jokioinen/Finland; Professor Dr. A. Finck, Institute of
      Plant Nutrition and Soil Science, Kiel/Federal Republic of
      Germany; Dr. A. Faber, Fertilizer and Soil Research
      Institute, Pulawy/Poland; Dr. L. Tiganas (Mrs.), Research
      Institute for Soil Science and Agrochemistry,
      Bucarest/Romania; Dr. J. Karlinger, Ministry of
      Agriculture and Food, Budapest/Hungary; Dr. N. Ulgen,
      Director, Soil and Fertilizer Research Institute,
      Ankara/Turkey; Dr. I. Aasen, Department of Soil Fertility
      and Management, Agricultural University, Aas/Norway;
      Dr. J. Sippola, Institute of Soil Science, Agricultural
      Research Centre, Jokioinen/Finland; Mr. J. Dissing
      Nielsen, State Laboratory for Soil and Crop Research,
      Lyngby/Denmark; Dr. J.-P. Quinche, Agronomic Research
      Station Changins, Nyon/Switzerland. Professor
      Dr. W. Podlesak and Professor Dr. O. Krause, Institute of
      Plant Nutrition of the Academy of Agricultural Sciences,
      Jena/Germany Democratic Republic.
      A report from the Byelorussian SSR (I.M. Bogdevich) and
      another report from the USSR (W.P. Tolstousov) were
      submitted, but not presented and discussed at the
      symposium due to the absence of the authors.

7.   Furthermore, the following additional report was presented:

Professor Dr. S. Manojlovic; Boron in the Chernozem of the
Vojvodina Province and its Effect on Sugarbeet Yields; Faculty
of Agriculture, Novi Sad/Yugoslavia.

The participants from the Netherlands and from Morocco as well as
other participants from the Federal Republic of Germany and the
United Kingdom also reported on the experiences and problems
concerning micronutrients in their respective countries.

8.   According to a recommendation by the previous symposium in this
series on soil fertility and fertilizers (held in 1983), the meeting
did not deal specifically with the toxic effects of heavy metals.
The papers presented and the discussions dealt with the following
main points:

I.   Utilization of secondary nutrients in agriculture (calcium,
magnesium, sulphur, etc.)

9.   Three of the reports under this sub-item dealt with one or
several secondary nutrients and reported on the relevant experiences
in their respective countries while a fourth report dealt with both
secondary and trace elements.

10.   The Mg deficiency in relation to grassland production has been
the subject of increasing attention in Sweden. Grass tetany,
particularly of high-yielding dairy cows, was the most well-known Mg
related health problem of animals originating in the Mg status of
the pasture crop. Mg deficiency was rising as a result of farming
systems without livestock production and was enhanced by the purity
of chemical fertilizers, particularly on sandy soils. Research
results pointed also to various other factors that influenced the
availability of Mg, e.g. K was found to depress the uptake of Mg;
the water content of the soil was of influence as was the
composition of the grassland (grass-clover ratio), the cutting (or
grazing) time or the level of N fertilization. Fertilization of
grassland with 50 kg of Mg/ha has increased the Mg concentration,
alternatively Mg can be supplied to dairy cows, both methods being
effective in reducing the frequency of grass tetany.

11.   A report from the United Kingdom dealt with the occurrence of S
deficiency, a phenomenon of a fairly recent nature. While earlier
investigations did not indicate sulphur deficiencies in most areas
of the United Kingdom, more recent studies provided some evidence
for response to S in rural areas of England, Wales and Scotland
where S pollution was low. The trend towards S deficiency will
probably continue as a result of declining sulphur dioxide
pollution, a lack in the use of sulphur containing fertilizers (e.g.
ammonium sulfate) and with further rising crop yields.

12. A report from the USSR showed that the intensification of crop
production increased the export of calcium and magnesium from the
soils. To this end, and also in view of the wide importance of
highly acid sod-podzolic soils in the country, liming with dolomitic
meals has become a common practice. The Mg supplies in soils and to
crops could be improved through the use of slags from the
metallurgical industries. The use of ashes from electrical power
stations in the form of ground material with a particle size of
below 0.25 mm could also enrich soils in Ca and Mg.

13. A report from Poland identified as the most urgent soil
fertility problem in regard to secondary and trace elements the
increased supply of lime, magnesium and boron, particularly on light
and organic soils. Furthermore, there was a need for initiating the
large-scale manufacturing of fertilizers containing copper and
molybedenum. Experiments have indicated Cu and Co deficiencies as a
big problem in feeds based on roughage from the grassland of the
country, occasionally also a lack of Mg and Zn. In these cases,
feeds would require supplementation of the deficient
micronutrients. The most economic way of applying secondary and
trace elements to crops would be leaf and seed application.

II. <u>Utilization of trace elements (micronutrients) in agriculture</u>
(boron, iron, zinc, copper, manganese, molybdenum, etc.)

14. A total of thirteen reports were presented under this
sub-item. One of these reports, presented by an expert from
Finland, reviewed the world-wide situation of micronutrients in
soils, based on samples taken in thirty countries under an FAO
project. As could be expected, these world-wide samples showed
extremely wide variations. For all of the six micronutrients
studied (B, Cu, Mo, Fe, Mn and Zn), soil pH seemed to play a
principal role in regulating their availability to plants.
Typically, countries with a low B, Cu and Mo status and with a high
Fe, Mn and Zn level had strongly to moderately acid soils. In
countries with alkaline soils the situation was the opposite.
Similar relationships were observed for the electrical conductivity
of soils. Other factors associated with micronutrient availability
in soils were the organic matter content, the cation exchange
capacity (CEC) and texture.

15. A report from the Federal Republic of Germany dealt with the
manganese deficiency that has become apparent in the northern
regions of the country. This deficiency which has become a major
yield-limiting factor in cereals and other intensively cultivated
crops, has been brought about mainly by regular liming on naturally
acid soils. Plant analysis was the best method of analysing Mn

deficiency. Measures to remedy this deficiency were still unsatisfactory, particularly because Mn application to the soil or to the leaves had no lasting effect. The search for new concepts of solving this problem was necessary, e.g. by temporary soil acidification.

16. A report from Poland reported on field experiments aimed at determining the initial and residual effects of micronutrients applied in a crop rotation (sugar beet, maize, peas, winter wheat). The following rates of application were found to remain effective throughout a three-year rotation: B - 2kg/ha; Cu - 10 kg/ha and Mo - 0.5 kg/ha whereas Zn applied at 5 to 20 kg/ha only lasted for two successive crops.

17. A report from Romania reported on research done in this country on micronutrient deficiencies and their effects, particularly regarding zinc, molybdenum and boron. The application of micronutrient fertilizers was found appropriate only in combination with a range of cultivation practices aiming at optimizing the use of available micronutrients in the soil. With limited supplies of micronutrient fertilizers, their use ought to be concentrated in seed production.

18. A report from Hungary noted that Cu and Zn deficiency was most frequent in this country, notably for winter cereals. Occasionally B deficiency could be observed in sunflowers. Furthermore, there was an indication of a certain P-Zn antagonism in soils. Both soil and foliar application of micronutrients were practiced, with a preference for the latter.

19. A report from Turkey reviewed the micronutrient situation of soils in this country. The most generalized and important deficiency concernced Zn and Fe whereas Mn and Cu were deficient only in some locations. B was available at normal to high rates in the soils and the content of Mo was normal. Zn deficiency was e.g. common in the citrus growing areas of southern Anatolia and Zn fertilization was common practice in these locations. In central Anatolia and in the Mediterranean regions lime induced chlorosis was widespread on calcareous soils and a good response to Fe and Zn application could be observed.

20. A report from Norway dealt with the micronutrient problems in this country. Zn was deficient on calcareous or on heavily limed soils (pH > 7) and an increasing deficiency was shown for vegetables, fruit crops, cereals, potatoes and also in grasses. Foliar application was recommended and practiced, leading also to an accumulation of residues in soils, thus increasing the availability

to future crops. Other common deficiencies observed were: Cu – on
peat and sandy soils; B – on peat and sandy soils and also on clay
soils; Fe – only on acid peat soils; Mn – under conditions of a
high soil pH and with a light textured soil structure.

21. A report from Finland stated that Finnish soils generally had a
low to moderate micronutrient content. The differences observed
between the various soil types were largely related to differences
in the texture. Emphasis had to be put on the micronutrients
available to plants. The most important deficiencies in the country
were concerning B and Cu whereas Mn, Mo and Zn deficiencies were
less frequent. Since B deficiency was common all over the country,
all commercial NPK fertilizers contained a B supplement. Zn was
supplemented mainly because of the requirements of the cattle in its
feed. Averages dosages provided through the supplementation of NPK
fertilizers were for B – 350 g; Cu – 300 g and for Mn – 200 g per
ha. The latest addition was introduced in 1984 and concerned
selenium in order to increase the Se intake of cattle and also of
the human population.

22. A report from Denmark investigated the pollution from heavy
metals through fertilizers, sludges and air pollution. Experiments
had been conducted with a view to study the relations between the
mineralogical composition of the soil and its Cu absorption. This
absorption was found to be confined particularly to clay, silt and
humus.

23. A report from Switzerland stated that an intensified nitrogen
fertilization did practically not influence the micronutrient
content (Fe, Mn, Cu, Zn, Li and B) of wheat and potatoes. However,
a reduced K and Mg content of potato tubers was found at a nitrogen
fertilization of 240 kg N/ha.

24. A report from the German Democratic Republic stated a low
availability of Cu, B and Mo in this country. High rates of
application were necessary due to the fixing of Cu in soils and to
the leaching of B and Mo from soils. Therefore fertilization with B
and Cu was practiced.

25. Another report from the German Democratic Republic dealt with
the factors that influenced the dosages of micronutrient
fertilization, e.g. the micronutrient contents of specific soil
types, the soil pH and the specific crop yield levels. Regarding
the application of micronutrient fertilizers there was a preference
for foliage application, but also seed treatment and band placement
were practiced.

26. A report from Yugoslavia dealt with the effects of boron on sugarbeet yields on chernozem soils. Experiments carried out under production conditions indicated some relationship between B available in soils and the yield parameters of sugarbeet only in certain years. Normally, the chernozem soils investigated in the Vojvodina Province appeared to be sufficiently rich in B so as to not require any specific B fertilization for sugar beets.

Conclusions and Recommendations

27. The Symposium formulated the following general conclusions and recommendations:

(i) There were considerable variations in the soil status of secondary and trace elements and even more so in the availability of these elements to plants. These variations were due to differences in the characteristics of the various soil types, the main factors being the soil pH and the texture of soils. There were a number of other factors, however, which were tending to change the situation over time, particularly the type of crops produced and their yield levels, the type of NPK fertilizers and also of organic fertilizers applied and the input from the air. Thus significant differences in the soil availability of certain trace elements could be observed in the vicinity of cities and big industrial agglomerations. Certain practices such as liming were also changing the availability of certain trace elements, e.g. Mn, to plants over longer periods of time. Most of the speakers tended to foresee an increasing importance of the balancing of the micronutrient availability for future crop production under the conditions of an intensive agricultural production.

(ii) Regarding secondary elements, lime requirements and Mg deficiencies were well known in many countries and regions and liming was therefore a common practice. A particular problem was the deficiency of Mg in grass on sandy soils which could cause health problems in dairy cattle. In some instances, rising yields, the use of sulphur-free fertilizers and decreasing air pollution might also call for increasing attention to the sulphur availability from soils.

(iii) concerning trace elements (micronutrients), B, Mn and Cu seemed to be the most important deficiencies in Europe; Mo, Zn and Fe were also giving rise to problems, but under more special circumstances. In Turkey, Morocco and other Mediterranean countries, Zn, Fe and Mn were most important. There was a rather general agreement on the need for the

application of the deficient micronutrients, preferably on the
basis of analysis. However, commercial supplies of
micronutrient fertilizers and/or micronutrient supplementation
of the basic NPK fertilizers were not always available in
sufficient quantities. For arable crops, leaf application of
Mn and possibly Zn was mostly seen as the most economic and
efficient way of fertilization, whereas for Cu soil
incorporation was preferred and for B and Zn both methods were
satisfactory. Seed treatment with Mo seemed preferable to
foliar application and more consideration should be given to Mo
and Zn treatment of the seed crop. For perennial fruit crops
foliar application was normally superior. The long term effect
of soil application was not always sufficient as yet and
further research into more efficient methods was suggested.
Where trace element deficiencies occurred in grassland, soil
application of Cu, Co and Se were effective, but direct
supplementation of fodder may also be used.

(iv) Many different methods of soil analysis for the
determination of available nutrients and other soil parameters
were currently being used in ECE countries. This diversity of
the method applied made any comparison of research results
difficult and there seemed therefore to be a need for
correlation and finally a general harmonization of these
extraction methods as well as for the establishment of standard
conversion factors in comparing research results. A possible
forum for undertaking this task could be the European
Co-operative Research Network on Trace Elements of FAO.

28. The participants of the Symposium considered the material
presented to be of general interest both in European and in
non-European countries. They therefore recommended the reports to
be made available for wider distribution, preferably by issuing a
publication of the complete proceedings.

29. The Symposium was informed that the ECE Committee on
Agricultural Problems, together with FAO, usually organized a
symposium in the field of soil fertility and fertilizers at
four-year intervals. For the attention of the Committee on
Agricultural Problems, the Symposium suggested the following
topic(s) for another FAO/ECE symposium to be held probably in 1991:

(i)   Soil and fertilizer requirements for the production of high-quality food and fodder (i.e. high in organic and inorganic essential components, but low in harmful substances).

(ii) The impact of environmental pollution on agriculture with special attention to the basis for current regulations in different countries.

Adoption of the Report

30. The Symposium adopted the present report.

RAPPORT DU COLLOQUE SUR L'UTILISATION
DES ELEMENTS SECONDAIRES ET DES
OLIGO-ELEMENTS DANS L'AGRICULTURE

Introduction

1. Le Colloque s'est tenu au Palais des Nations à Genève du 12 au 16 janvier 1987. Y ont participé 26 experts représentant les gouvernements, des instituts de recherche et l'industrie des pays suivants : Allemagne, République fédérale d'; Belgique; Danemark; Finlande; Hongrie; Norvège; Pays-Bas; Pologne; Portugal, République démocratique allemande; Roumanie; Royaume-Uni; Suède; Suisse; Turquie, Union des Républiques socialistes soviétiques et Yougoslavie. Le Maroc y a participé en application de l'article 11 du mandat de la Commission économique des Nations Unies pour l'Europe.

2. Des représentants des organisations non gouvernementales suivantes ont participé au Colloque : Centre international des engrais chimiques (CIEC); Institut international de la potasse.

3. La réunion a été ouverte par le Directeur de la Division FAO/CEE de l'agriculture et du bois.

Adoption de l'ordre du jour

4. L'ordre du jour provisoire (AGRI/SEM.21/1/Add.1) a été adopté.

Election du Bureau

5. M. A. Finck (République fédérale d'Allemagne) a été élu président du Colloque, et M. H. Gembarzewski (Pologne) vice-président.

Programme

6. Le programme du Colloque était composé comme suit :

I. <u>Utilisation des nutriments secondaires dans l'agriculture (calcium, magnésium, soufre, etc.)</u>

Rapporteurs : M. L.-G. Nilsson, Université suédoise des sciences agricoles, Uppsala (Suède); M. D. Hewgill, Ministère de l'agriculture, des pêcheries et de l'alimentation, Newcastle upon Tyne (Royaume-Uni); M. I.A. Chilnikov, directeur du laboratoire de l'Institut national de recherche pour les engrais et les sciences du sol, Moscou (URSS) et M. Gembarzewski, Institut de recherche sur les engrais et les sols, Wroclaw (Pologne).

II. <u>Utilisation des oligo-éléments (micronutriments) dans l'agriculture (bore, fer, zinc, cuivre, manganèse, molybdène, etc.)</u>

Rapporteurs : M. M. Sillanpää, Directeur de l'Institut de pédologie, Centre de recherche agricole, Jokioinen (Finlande); M. A. Finck, Institut de recherche en matière de nutrition des plantes et de science du sol, Kiel (République fédérale d'Allemagne); M. A. Faber, Institut de recherche sur les engrais et les sols, Pulawy (Pologne); Mme L. Tiganas, Institut de recherche pour la pédologie et l'agrochimie, Bucarest (Roumanie); M. J. Karlinger, Ministère de l'agriculture et de l'alimentation, Budapest (Hongrie); M. N. Ulgen, Directeur de l'Institut de recherche sur les sols et les engrais, Ankara (Turquie); M. I. Assen, Département de l'exploitation et de la fertilité des sols de l'Université agricole, Aas (Norvège). M. J. Sippola, Institut de pédologie, Centre de recherche agricole, Jokioinen (Finlande); M. J. Dissing Nielsen, Laboratoire national de recherche sur les sols et les cultures, Lyngby (Danemark); M. J.P. Quinche, Station fédérale de recherches agronomiques, Changins, Nyon (Suisse); M. W. Podlesack, Institut de recherche en matière de nutrition des plantes de l'Académie des sciences agricoles, Iéna (République démocratique allemande), et M. O. Krause, Institut de recherche en matière de nutrition des plantes de l'Académie des sciences agricoles, Iéna (République démocratique allemande).

Un rapport de la RSS de Biélorussie (I.M. Bogdevitch) et un rapport de l'URSS (W.P. Tolstouzov) ont été soumis, mais, leurs auteurs étant absents, ils n'ont été ni présentés ni examinés au Colloque.

7.   Le rapport additionnel suivant a été présenté :

M. S. Manojlovic : Le bore dans la zone du tchernoziom de la province de Vojvodine et ses effets sur le rendement de la betterave à sucre; Faculté d'agriculture, Novi Sad (Yougoslavie).

Les participants des Pays-Bas et du Maroc, ainsi que ceux de la République fédérale d'Allemagne et du Royaume-Uni, ont aussi fait part de l'expérience et des problèmes concernant les micronutriments dans leurs pays respectifs.

8.   Conformément à une recommandation sur la fertilité du sol et les engrais, formulée au colloque précédent dans cette série (1983), les participants n'ont pas traité spécialement des effets toxiques des métaux lourds. Les documents présentés et le débat ont porté sur les principales questions suivantes :

I.   Utilisation des nutriments secondaires dans l'agriculture (calcium, magnésium, soufre)

9.   Trois des quatre rapports présentés au titre de cette question traitaient d'un ou de plusieurs nutriments secondaires, relatant l'expérience acquise dans les pays considérés, et le quatrième portait à la fois sur les éléments secondaires et les oligo-éléments.

10. Les effets de la carence en magnésium sur la production prairiale ont fait l'objet d'une attention accrue en Suède. Chez les animaux, et surtout chez la vache laitière à rendement élevé, la tétanie d'herbage était, parmi les problèmes sanitaires liés au magnésium et dus à la teneur en magnésium du fourrage, celui qui était le mieux connu. La carence en magnésium devenait de plus en plus marquée avec le développement des techniques de production agricole sans élevage et était encore aggravée par l'épandage d'engrais chimiques purs, surtout sur sols sableux. Il ressortait aussi des travaux de recherche que divers autres facteurs influaient sur les disponibilités en magnésium : c'est ainsi que le potassium diminuait l'absorption de magnésium, que la teneur en eau du sol avait aussi un rôle, de même que la composition de l'herbage (rapport graminées-trèfle), la période de coupe (ou de pâture) ou l'importance de l'apport en azote. La fertilisation des herbages à raison de 50 kg de Mg/ha permettait d'accroître la teneur en magnésium, et par ailleurs du magnésium pouvait être administré aux vaches laitières, ces deux méthodes étant efficaces pour réduire la fréquence de la tétanie d'herbage.

11. Un rapport du Royaume-Uni portait sur les cas de déficience en soufre, phénomène assez récent. Les enquêtes précédentes n'avaient révélé aucune déficience dans la plupart des régions du Royaume-Uni, mais des études plus récentes ont montré qu'il y avait une réaction positive au soufre dans certaines zones rurales d'Angleterre, du pays de Galles et d'Ecosse où la pollution par cette substance était faible. La tendance à une déficience en soufre va sans doute se maintenir, en raison d'une diminution de la pollution par le dioxyde de soufre, de la non-utilisation d'engrais contenant du soufre (par exemple le sulfate d'ammonium) et d'un nouvel accroissement des rendements.

12. Il ressortait d'un rapport de l'URSS que l'intensification du rendement des cultures accroissait les pertes du sol en calcium et en magnésium. Pour y remédier et compte tenu aussi de la part importante des sols gazonnés podzoliques très acides dans le pays, le chaulage au moyen de farines de dolomite était aujourd'hui de pratique courante. Les apports en magnésium dans le sol et les cultures pourraient être améliorés en recourant aux scories produites par les industries métallurgiques. L'emploi de cendres de centrales électriques sous forme de matériaux broyés de granulométrie inférieure à 0,25 mm pourrait aussi enrichir les sols en calcium et en magnésium.

13. Un rapport de la Pologne indiquait que le problème de fertilité du sol le plus urgent du point de vue des éléments secondaires et des oligo-éléments concernait l'augmentation des apports de chaux, de magnésium et de bore, surtout sur les sols légers et organiques. Il était nécessaire aussi d'entreprendre la production à grande échelle d'engrais contenant du cuivre et du molybdène. Des expériences ont montré que les carences en cuivre et en cobalt constituaient un problème important pour les fourrages grossiers provenant des prairies polonaises, tout comme parfois les carences en magnésium et en zinc. Dans ces cas, les rations devaient être complétées par un apport en micronutriments déficitaires. Le mode le plus économique d'apport d'éléments secondaires et d'oligo-éléments dans les cultures consisterait à les appliquer sur les feuilles et les graines.

II. Utilisation d'oligo-éléments (micronutriments) dans l'agriculture (bore, fer, zinc, cuivre, manganèse, molybdène, etc.)

14. Au total 13 rapports ont été présentés au titre de cette question. Dans l'un d'eux, présenté par un expert de la Finlande, la situation mondiale des micronutriments dans les sols était examinée à partir d'échantillons prélevés dans trente pays dans le cadre d'un projet FAO. Comme on pouvait s'y attendre, ces

échantillons, choisis dans le monde entier, présentaient entre eux
de très grandes différences. Pour les six micronutriments étudiés
(B, Cu, Mo, Fe, Mn et Zn), c'est le pH du sol qui paraissait avoir
un rôle essentiel dans la régulation de leur disponibilité pour les
plantes. Ainsi, dans les pays où les sols avaient une faible teneur
en bore, en cuivre et en molybdène et une teneur élevée en fer,
maganèse et zinc, ces sols étaient de fortement à modérément
acides. Dans les pays à sols alcalins, la situation était inverse.
Des relations semblables ont été observées en ce qui concerne la
conductibilité électrique des sols. La teneur en matières
organiques, la capacité d'échange cationique et la texture étaient
aussi des facteurs liés à la disponibilité des micronutriments dans
les sols.

15. Un rapport de la République fédérale d'Allemagne portait sur la
carence en manganèse qui s'est manifestée dans les régions
septentrionales du pays. Cette carence, qui est devenue le
principal facteur limitatif du rendement en céréales et autres
produits de culture intensive, a été provoquée surtout par le
chaulage périodique de sols naturellement acides. L'analyse des
plantes était la meilleure méthode à appliquer pour analyser la
carence en manganèse. Les mesures prises pour y remédier restaient
peu satisfaisantes, en particulier parce que l'application de
manganèse sur le sol ou sur les feuilles n'avait pas d'effet
durable. Il fallait donc chercher de nouvelles solutions pour
résoudre ce problème, par exemple l'acidification temporaire du sol.

16. Un rapport de la Pologne était consacré aux expériences faites
sur le terrain pour déterminer les effets initiaux et résiduels des
micronutriments appliqués dans un système de rotation des cultures
(betterave sucrière, maïs, pois, blé d'hiver). Les taux
d'application suivants se sont révélés efficaces sur une rotation de
trois ans : B – 2 kg/ha, Cu – 10 kg/ha et Mo – 0,5 kg/ha, alors que
le zinc appliqué à raison de 5 à 20 kg/ha n'a produit d'effet que
pendant deux récoltes successives.

17. Un rapport de la Roumanie exposait les recherches faites dans
ce pays sur les carences en micronutriments et leurs effets, en
particulier en ce qui concerne le zinc, le molybdène et le bore.
L'application d'engrais contenant des micronutriments s'est révélée
appropriée seulement en combinaison avec un ensemble de pratiques
culturales visant à optimiser l'utilisation des micronutriments
présents dans le sol. Les quantités d'engrais aux micronutriments
étant limitées, il convenait de les utiliser en premier lieu pour la
production de semences.

18. Selon un rapport de la Hongrie, c'était la carence en cuivre et
en zinc qui était la plus fréquente dans ce pays, notamment pour les
céréales d'hiver. On observait aussi parfois une carence en bore
dans les tournesols. En outre, on notait des signes d'un certain
antagonisme P-Zn dans les sols. L'application de micronutriments
était pratiquée tant sur les sols que sur les feuilles, avec une
préférence pour celles-ci.

19. Un rapport de la Turquie passait en revue la situation des
micronutriments présents dans les sols de ce pays. La carence la
plus importante et la plus répandue concernait le zinc et le fer,
alors que le manganèse et le cuivre manquaient seulement en certains
endroits. Le bore était présent en quantités normales à élevées
dans les sols, et la teneur en molybdène était normale. La carence
en zinc était par exemple courante dans les régions consacrées à la
culture des agrumes de l'Anatolie méridionale et l'application
d'engrais contenant du zinc était couramment pratiquée dans ces
régions. En Anatolie centrale et dans les régions méditarranéennes,
la chlorose provoquée par le chaulage était répandue sur les sols
calcaires, et on a observé une bonne réaction à l'application de fer
et de zinc.

20. Un rapport de la Norvège portait sur les problèmes posés par
les micronutriments dans ce pays. Le zinc manquait dans les sols
calcaires ou fortement chaulés (pH>7) et une carence croissante se
révélait pour les légumes, les fruits, les céréales, les pommes de
terre, ainsi que les herbages. L'application sur les feuilles était
recommandée et pratiquée, et entraînait une accumulation de résidus
dans les sols, ce qui accroissait la disponibilité pour les récoltes
futures. Les autres carences couramment constatées étaient les
suivantes : Cu - sur les sols tourbeux et sablonneux; B - sur les
sols tourbeux et sablonneux, ainsi que sur les sols argileux;
Fe - seulement sur les sols tourbeux acides; Mn - en cas de pH élevé
et dans les sols meubles.

21. Un rapport de la Finlande indiquait que les sols du pays
avaient en général une teneur faible à modérée en micronutriments.
Les différences observées entre les divers types de sols étaient
largement liées aux différences de texture. Il fallait mettre
l'accent sur les micronutriments à disposition des plantes. Les
carences les plus importantes concernaient le bore, et le cuivre,
alors que celles de manganèse, de molybdène et de zinc étaient moins
fréquentes. Comme la carence en bore était générale dans le pays,
tous les engrais NPK qu'on trouvait dans le commerce en contenaient
une adjonction. Le zinc était ajouté principalement pour répondre
aux besoins alimentaires du bétail. Les dosages moyens assurés par
adjonction dans les engrais NPK étaient les suivants : B - 350 g;
Cu - 300 g et Mn - 200 g par ha. L'adjonction la plus récente
remontait à 1984; il s'agissait de sélénium ajouté en vue d'en
accroître l'absorption par le bétail, ainsi que par la population.

16

22. Un rapport du Danemark examinait la pollution provoquée par les métaux lourds, due à l'épandage d'engrais et de boues et à la pollution de l'air. Des expériences ont été faites pour étudier les rapports entre la composition minéralogique du sol et son absorption de cuivre. Cette absorption s'est révélée limitée surtout à l'argile, au limon et à l'humus.

23. Un rapport de la Suisse indiquait qu'une application intensifiée d'engrais azotés n'avait quasiment aucune influence sur la teneur en micronutriments (Fe, Mn, Cu, Zn, Li et B) du blé et des pommes de terre. Toutefois, une teneur réduite en potassium et en magnésium dans les tubercules de pommes de terre a été constatée à 240 kg de N/ha.

24. Un rapport de la République démocratique allemande indiquait une faible disponibilité de cuivre, de bore et de molybdène dans le pays. Des taux d'application élevés étaient nécessaires en raison de la fixation du cuivre dans les sols et du lessivage du bore et du molybdène. C'est pourquoi, on pratiquait l'épandage d'engrais contenant du bore et du cuivre.

25. Un autre rapport de la République démocratique allemande portait sur les facteurs influant sur les dosages d'engrais contenant des micronutriments, par exemple la teneur en micronutriments de certains types de sols, le pH du sol et les niveaux de rendement des cultures. En ce qui concerne l'application d'engrais contenant des micronutriments, la préférence était donnée à l'application sur les feuilles, mais on pratiquait aussi le traitement des semences et la localisation en bande.

26. Un rapport de la Yougoslavie portait sur les effets du bore sur le rendement de la betterave sucrière cultivée en tchernoziom. Les expériences réalisées en conditions de production faisaient apparaître un rapport entre le bore présent dans les sols et les paramètres de rendement de la betterave sucrière, mais seulement pour certaines années. Normalement, le tchernoziom étudié dans la province de Vojvodine semblait suffisamment riche en bore pour qu'il ne soit pas nécessaire d'appliquer un engrais quelconque contenant du bore pour la culture des betteraves sucrières.

Conclusions et recommandations

27. Le Colloque a formulé les conclusions et recommandations générales suivantes :

i) Des différences considérables ont été constatées dans la teneur des sols en éléments secondaires et oligo-éléments, ces différences étant encore plus marquées eu égard à la disponibilité de ces éléments pour les plantes. Ces différences tenaient à la variété des caractéristiques des divers types de sols, les facteurs principaux étant le pH et la texture du sol. Cependant, plusieurs autres facteurs tendaient à modifier la situation dans le temps, et en particulier, le type de culture et les rendements, les types d'engrais NPK et d'engrais organiques utilisés ainsi que les apports atmosphériques. On pouvait donc observer des différences importantes dans la disponibilité de certains oligo-éléments dans le sol au voisinage des villes et des grandes agglomérations industrielles. Certaines méthodes, telles que le chaulage, modifiaient aussi la disponibilité de certains oligo-éléments, par exemple le manganèse, pour les plantes pendant de longues périodes. La plupartdes orateurs ont en général estimé qu'il serait de plus en plus important pour les récoltes futures, dans une situation de production agricole intensive, de veiller à assurer l'équilibre entre les micronutriments disponibles.

ii) En ce qui concerne les éléments secondaires, les besoins en chaux et les carences en magnésium étaient bien connus dans de nombreux pays et régions, où le chaulage était donc une méthode couramment appliquée. Un problème particulier était posé par la carence en magnésium des herbages sur sol sableux, qui risquait de provoquer des problèmes sanitaires chez les vaches laitières. Dans certains cas, l'accroissement des rendements, l'utilisation d'engrais NPK sans soufre et la diminution de la pollution de l'air pourraient aussi conduire à accorder davantage d'attention à la disponibilité du soufre dans les sols.

iii) En ce qui concerne les oligo-éléments (micronutriments), les carences en bore, manganèse et cuivre paraissaient les plus importantes en Europe. Le molybdène, le zinc et le fer donnaient lieu aussi à des problèmes, mais dans des cas plus particuliers. En Turquie, au Maroc et dans d'autres pays méditerranéens, c'étaient les carences en zinc, fer et manganèse qui étaient les plus importantes.

La plupart des participants ont estimé qu'il était nécessaire de faire des apports en micronutriments déficitaires, de préférence après analyse. Cependant, les engrais à micronutriments et/ou les micronutriments destinés à compléter les engrais de base NPK n'étaient pas toujours disponibles en quantités suffisantes dans le commerce. Pour les cultures de labour, l'application de manganèse, et éventuellement de zinc, sur les feuilles était le plus souvent considérée comme le mode de fertilisation le plus économique et le plus efficace, alors que pour le cuivre, la préférence était donnée à l'adjonction dans le sol et, pour le bore et le zinc, les deux méthodes étaient jugées satisfaisantes. Le traitement des semences au molybdène semblait être une meilleure méthode que l'application sur les feuilles, et il faudrait s'intéresser davantage au traitement des plantes à graines au molybdène et au zinc. Pour les cultures fruitières vivaces, c'était normalement l'application sur les feuilles qui donnait les meilleurs résultats. On ne connaissait pas encore suffisamment les conséquences à long terme de l'apport dans le sol, et il a été suggéré de poursuivre les travaux de recherche pour trouver des méthodes plus efficaces. En cas de carences des herbages en oligo-éléments, l'apport de cuivre, de cobalt et de sélénium était efficace, mais il était possible aussi d'apporter un complément direct au fourrage.

iv) De nombreuses méthodes différentes d'analyse du sol étaient actuellement utilisées dans les pays de la CEE pour déterminer les nutriments disponibles ainsi que d'autres paramètres du sol. La diversité des méthodes appliquées rendait difficile toute comparaison des résultats de la recherche et il paraissait donc nécessaire d'établir, aux fins d'harmonisation, une corrélation entre toutes les méthodes d'extraction et de définir des facteurs uniformes de conversion pour comparer les résultats de la recherche. Un organe approprié pour entreprendre ce travail pourrait être le Réseau européen de recherche en coopération sur les oligo-éléments de la FAO.

28. Les participants au Colloque ont estimé que les documents présentés étaient d'intérêt général, aussi bien pour les pays européens que pour des pays situés hors d'Europe. Ils ont donc recommandé d'en prévoir une distribution élargie, de préférence sous forme de publication qui réunirait l'ensemble des Actes du Colloque.

29.   Les participants ont été informés que le Comité des problèmes agricoles de la CEE, agissant conjointement avec la FAO, organisait généralement un colloque consacré à la fertilité du sol et aux engrais tous les quatre ans.   A son intention, ils ont proposé les thèmes ci-après pour le colloque FAO/CEE qui aura probablement lieu en 1991 :

    i)   Besoins en sols et engrais pour la production d'aliments et de fourrages de haute qualité (c'est-à-dire riches en éléments organiques et inorganiques essentiels, mais contenant peu de substances nocives).

    ii)   L'impact de la pollution environnementale sur l'agriculture, eu égard en particulier aux éléments de base des réglementations en vigueur dans les différents pays.

30.   Les participants au Colloque ont adopté le présent rapport.

MAGNESIUM IN GRASSLAND PRODUCTION

Mr. L.G. Nilsson, Department of Soil
Science, Swedish University of
Agricultural Sciences, Uppsala

INTRODUCTION

Magnesium is a plant nutrient which has received increasing attention in Swedish agriculture during the last decades. A reason for this is the intensified cultivation with a high level of crop production. A high yield involves great demand on a comprehensive and well balanced supply of plant nutrients. One other factor influencing negatively the magnesium status in the soils is the great number of farms without livestock production. The generally favourable effect of farmyard manure on the plant nutrient situation in the soil is well documented.

Previously Mg was applied unwittingly as an impurity along with other fertilizers. The high purity of fertilizers used at present means, however, that this source of Mg application to the soil no longer exists. For this reason and because of the other factors mentioned Mg deficiency in crop plants is becoming more frequent and application of Mg-containing fertilizers is now common.

Concerning the supply of Mg in the soil it can generally be noted that deficiencies more often occur in sandy soils than in soils with a high content of clay. The principal reasons for this are that the sandy soils as a rule are lacking Mg in nature and that losses by leaching can amount to 50 Kg of Mg per ha in these coarse textured soils.

The status of Mg in the soils in Sweden is determined according to the AL-method (extraction solution: ammonium lactate, pH = 3.75) and the critical values for deficit range from 2 to 10 mg per 100 g of soil. The low values refer to sandy soils and the high ones to clay soils rich in potassium. This adaptation of the critical value to the level of potassium in soil depends on the antagonism between potassium and magnesium. Our experiences up to now indicate that the quotient of K-AL and Mg-Al ought to be about 2.

Consequently, the use of fertilizers over a long period has been characterized by unilateral application of the main plant nutrients nitrogen, phosphorus and potassium. This has in many cases resulted in negative effects of potassium applied and an analysis reveals generally that this depends on a deficiency of magnesium in the soil, in other words on a unfavourable balance between the two elements (Fig. 1).

The Mg uptake of plants is about of the same size as for phosphorus or 10-30 kg P per ha. Already at the beginning of this century it was observed that Mg was an important constituent of the chlorophyll. It is estimated that about 20% of the magnesium taken up by the plants becomes an integral part of the chlorophyll and the rest takes part in many other important physiological functions in the plant. The relationship between Mg and the chlorophyll content and between Mg and protein synthesis is demonstrated in Table 1 (Michael, 1941).

Application of dolomite to grassland has produced a significant increase in yield under the condition of a low pH value in association with a low level of Mg in the soil (Baerug, 1981). In current long-term field experiments with grass on sandy soils in Sweden yield increases of 10% were obtained in 1983 after an application of magnesium sulfate, despite a basic fertilization of 3000 kg of dolomite lime per hectare, corresponding to about 240 kg of Mg per hectare, in 1977. These results indicate that an application of Mg must be repeated with suitable intervals.

The significance of a suitable ratio of K to Mg in soil for crop production has been discussed. The depression of Mg uptake by plants from application of K is widely known and it has also been noted in many investigations that the most important single factor influencing the Mg uptake is the quantity of available K. In Figure.2 this relationship can be demonstrated in principle by a pasture experiment carried out in Austria (Gunhold, 1973). It is also interesting to note how the Mg content of the grass obviously affects the content of Mg in the blood serum of cattle. In the literature two critical levels are often mentioned, namely that the Mg content of blood serum should not be below 2 mg per 100 ml. If these requirements not are met, the cattle runs a risk of being subject to tetany.

In the Netherlands (Committee on Mineral Nutrition, 1973) an Mg content about 0.30% DM in pasture grass rich in crude protein and potassium is recommended at early stages . In older grass with a content of crude protein of less than 20% and with a potassium content below 3%, an Mg content of about 0.20% is acceptable. According to our own experience in Sweden a value of 0.15% Mg in DM can be satisfactory when feeding with hay.

If we examine the general critical value 0.2 mg Mg per 100 ml of blood serum it can be established that values of > 3.2 are excellent, 1.8-3.2 balanced and < 1.8 deficient (danger of tetany). (Simesen, 1977 and Meyer, 1963).

Grass tetany is certainly the most well-known of the different hypomagnesemias in ruminants. More than 75% of all cases happen during the first month after the start of the grazing season and especially 7-10 days after turning out. The tetany is often a combination of reduced Mg uptake from the feed and decreased Mg absorption in the digestive system. The net requirement of a cow depends on her daily milk yield and can be evaluated at 3 g for a dry cow and at 6 g for a milk production of 30 kg a day. These figures reveal that the risk of grass tetany is highest for high-producing cows.

During the 1950's grass tetany was a troublesome disease which mainly depended on the common way of turning the cows directly from the stable to spring pasture. During the first part of the 1960's there was intensive propaganda for giving supplementary Mg to dairy cows during the weeks before and after turning out to pasture. At the same time there was also propaganda for a gradual change-over feeding. When commercial protein feeds were widely used later on the situation was even better because the manufacturers supplied their protein feeds with Mg during spring and summer. All these procedures resulted in a greatly reduced frequency of grass tetany and this disease became relatively uncommon in Sweden. The reason why possibly the frequency of grass tetany could increase in the future is the increasing tendency to feed solely home-produced feedstuffs. However, most likely grass tetany will also be relatively uncommon in Sweden in the future.

Other factors associated with plant uptake and with the Mg absorption in the digestive system with a high probability of inducing grass tetany in ruminants are: a content of K higher than 2.5% DM, a high content of long-chain fatty acids, high concentrations of organic acids and the associated chelating capability, a high N content, etc. Concerning the high N content it has been established in the Netherlands in 1973 that the product of the content in % DM of crude protein and of K ought not to exceed 50. The most common way to characterize tetanigenic grass with regard to the content of minerals is the use of K/Ca +Mg ratios on an equivalent basis and this quotient ought not to exceed 2.2.

By application of Mg ($MgSO_4$) it is possible to increase the content of this element in the forage according to current field experiments in Sweden (Table 2). Trials carried out in Norway (Baerug, 1977) and in other places show corresponding results. These generally small increases in the Mg content in the forage are considered to be of great significance for feeding. It can also be established that the Mg content in many cases is higher in forage from the regrowth than from the first cutting, this probably being an effect of the reduced K content in these late harvests. The results presented also provide evidence of the negative effect of potassium on the Mg content of the forage. It is mentioned that it is very difficult to obtain favourable ratios between K and Mg in pasture when the K content exceeds 3%. According to results from field experiments in Sweden and Norway (Baerug, 1977), the maximum yield appears to be obtained at a content of K in the herbage of 2.0-2.5%, which means that any addition of K in order to reach higher contents is to be regarded as a luxury consumption and ought to be avoided.

Besides the direct fertilizing effect of K and Mg to the grassland there are also other factors affecting the mineral balance in the forage, e.g. the soil moisture, the botanical composition of the grass-clover relation, the stage of development and nitrogen fertilization.

Generally it can be mentioned that a higher water content in the soil promotes the uptake of K by plants more than do Mg and Ca which results in an increased quotient of K/Ca+Mg. Schuffelen (1954) established that increased soil water content increases K activity in soil solution rather than the activity of Mg.

Grunes (1967) found that ratios of K to Ca+Mg in rye grass were higher at lower than at higher temperatures. Fairbourn and Batchelder (1980) observed that forage Mg level increased with increasing soil temperature. These findings are well in accordance with agricultural practice namely that the risk of tetany is greater during wet and cold weather conditions in spring and in autumn.

Concerning the botanical composition it is well known that clover has much higher concentration of magnesium than grass. The varying Mg content in different ley plants has been investigated by Svanberg and Ekman (1946) and the results are presented in Table 3. Practical experience also demonstrates that tetany rarely occurs in pasture containing clover and when feeding with hay rich in clover. Both irrigation and a good status of K in the soil promote the share of clover in pastures and in leys and from this point of view these factors can also have a positive effect on Mg content in the forage, in contrast to the findings discussed earlier.

The influence of fertilization and of the stage of development on mineral composition has been studied in field trials carried out with timothy ley. The results are presented in Table 4 (Odelin, 1961). The quotient of K/Ca+Mg increases with an increasing rate of fertilizers at both stages of harvesting but it is much higher at an early stage than at a later stage of harvesting.

The effect of nitrogen fertilization on the mineral composition of forage must be looked upon as complicated not least with regard to the depressing effect of ammonium (like K) on the uptake of Mg by plants and to the positive influence of nitrate on the uptake of mineral elements in general. The effect of the nitrogen form on the cation composition of maize (Zea mays) can be demonstrated in principle with a greenhouse experiment (Table 5) conducted by Classen & Wilcox (1973). Ammonium-N decreased the percentage of K, Ca and Mg in the tissue and K also suppressed the Mg and Ca uptake by the plants. In a pot experiment with barley carried out in Denmark (Hansen, 1970) it was found that the quotients of K/Mg in the plants were higher throughout the growing season with nitrate-N added than with the application of ammonium-N. Baerug (1977) also found in Norwegian field experiments that application of nitrochalk (NH_4NO_3) increased the Mg content in the ley grass and it indicates that nitrate-N must have been the dominating form of N in the soil. Favourable conditions of nitrification in the soil have probably contributed to this situation. Another widely known effect of nitrogen fertilization is that it favours the development of grasses at the expense of clover and what this means to the content of Mg in the forage has been pointed out earlier.

To sum up, it can be emphasized that there are three factors in today's grassland production which have contributed to the current interest in the content of Mg in forage, i.e. the great increase in potassium fertilization, the diminished share of leguminous crops in the forage and the early stages of harvesting for ensiling and grazing.

REFERENCES

Baerug, R. 1977   Nitrogen, Kalium og svaveltil eng pa
                           Sör-Ostlandet. Norges landbrukshögskole.
                           Inst. for jordkultur Nr 91, 549-574.

Baerug, R. 1981   Magnesiumgjodsling till jordbruksvekster.
                           Forskning og forsok i landbruket, 32.
                           45-53.

Claassen, M.E. & Wilcox, G.E. 1973. Comparative Reduction of
                           Calcium and Magnesium Composition of Corn
                           Tissue by NH_4-N and K Fertilization J.
                           Paper No 5208 Agri Exp. St Purdue
                           University Lafayette, 521-522

Committee on Mineral Nutrition, 1973. Tracing and treating mineral
                           disorders in dairy cattle. Centre of
                           agricultural publishing and documentation,
                           Wageningen.

Fairbrown, M.L. & Batchelder, A.R. 1980. Factors influencing
                           magnesium in High Plains forage. J. of
                           Range Management, 33. 435-438.

Grunes, D.L. 1967.   Grass tetany of cattle as affected by
                           plant composition and organic acids.
                           Proc. Cornell Nutr. Conf. Food Mfr.
                           105-110.

Gunhold, P. 1973.   Die magnesiumversorgung österrichischer
                           Böden und ihre Bedeutung für die
                           Rinderhaltung. Symposium. Die Rolle von
                           magnesium and Schwefel in der
                           Pflanzenernährung.

Hansen, E.M. 1970.   Den kemiska sammensaetning af jordvaesken
                           og naeringsstofbalansen i planterne
                           gennem vaekstperioden ved tillforsel af
                           forskellige kvaelstofgodninger.
                           Licentiatafhandling ved Den kgl
                           Veterinaer- og Landbohojskole Kobenhavn.

Meyer, H. 1963.   Vet. Habil. Publ. Hannover, BRD

Michel. G. 1941.   Pflzern. Dg und Bdkde 25, 65-120.

Schuffeln, A.C. 1954.   The absorption of potassium by the plant.
                             Kali- Symposium. 169-181.

Svanberg, O. & Ekman, P. 1946.   Om magnesiumhalten i vegetationen
                             fran svenska jordar. Kungl.
                             Lantbruksakademiens Tidskr. 85, 54-99.

Odelien, M. 1961.   Kan gödsling förorsaka hypomagnesemi och
                             tetani hos nötkreatur? Växt-När-Nytt,
                             17:1. 1-8.

Table 1

Influence of the Mg Rate on the Content of Chlorophyll and on Protein-N (Michael, 1941)

Mg-fertilization	Chlorofyll (mg)	Protein-N (% DM)
Without	1.44	3.2
Limited	1.72	3.6
Ample	2.36	3.8

Table 2

Influence of Mg and K Fertilization to a Grass Ley on the Content of these Elements in the Crop

Application	% DM			
Application Element Kg ha-1	First harvest		Second harvest	
	Mg	K	Mg	K
Mg 0	0.16	2.02	0.15	0.87
20	0.17	2.24	0.15	0.75
80	0.20	2.15	0.16	0.80
160	0.22	2.07	0.19	0.76
K 0	0.26	0.88	0.21	0.52
80	0.20	1.84	0.18	0.73
160	0.16	2.41	0.15	0.81
640	0.12	3.35	0.10	1.13

Table 3

Content of Magnesium in Hay Samples in Sweden
(Svanberg and Ekman, 1964)

Samples Number	Plant	Mg, % DM Means	Variation
246	Red clover	0.34	0.16-0.70
18	Alsike clover	0.32	0.21-0.52
24	Alfa-Alfa	0.23	0.15-0.32
146	Timothy	0.13	0.06-0.26

Table 4

Quotient of K/Ca+Mg in Timothy Hay Harvested at Different Stages of Development; Average of 6 Field Experiments
(Odelien, 1961)

Complete fertilizer Kg ha $^{-1}$ K fertilizer (33%) Kg ha $^{-1}$	0 0	300 75	600 150	900 225
Harvest Heading Flowering	 1.39 1.15	 1.72 1.41	 1.99 1.48	 2.31 1.60

Table 5

Effect of N Form and K Rate on Growth and Composition of Maize Plants (Claassen and Wilcox, 1973)

Treatment		DM	% DM		
N-form*)	K ppm	g	K	Ca	Mg
NO$_3$	0	2.01	3.79	0.86	0.32
	50	2.24	4.51	0.79	0.27
	100	2.20	4.99	0.71	0.24
NH$_4$	0	1.00	2.98	0.67	0.20
	50	1.30	3.43	0.63	0.19
	100	1.89	3.66	0.69	0.18

*)   100 ppm

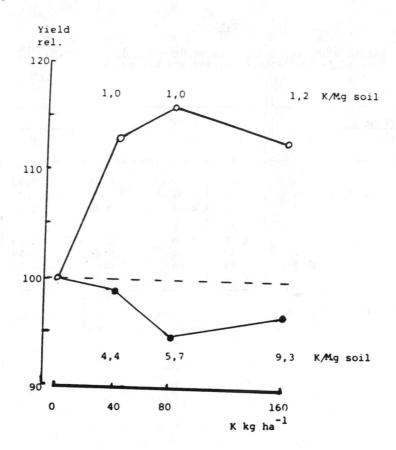

Fig. 1. Effect of K-fertilisation on yield at the same K-status
but at different Mg-status in the soil

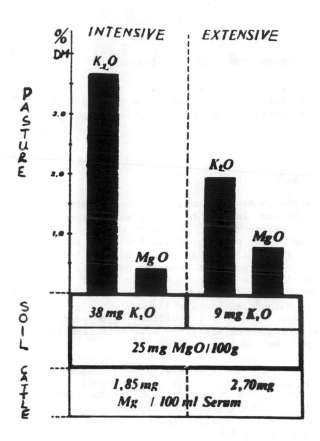

Fig. 2. K- and Mg relations at intensive and extensive grazing (Gunhold, 1973)

LE MAGNÉSIUM DANS LES HERBAGES

M. L.G. Nilsson,
Département de pédologie de l'Université
suédoise des sciences agricoles, Uppsala

RESUME

Le magnésium en tant que nutriment des plantes a conquis une place de plus en plus importante dans l'agriculture suédoise au cours des dernières décennies. En raison du grand nombre d'exploitations qui ne pratiquent pas l'élevage et de la grande pureté des engrais utilisés à l'heure actuelle, les carences en magnésium sont plus fréquentes dans les cultures; ces carences sont plus courantes dans les sols sablonneux que dans les sols argileux. Le rapport potassium/magnésium dans le sol, déterminé de préférence au moyen de la méthode au lactate d'ammonium, doit se situer aux environs de 2 pour maintenir l'équilibre du rapport K : Mg dans les plantes. L'application de magnésium doit être répétée à certains intervalles pour éviter les carences.

L'application d'engrais potassiques doit être limitée de manière que la teneur en potassium de la matière sèche des plantes n'excède pas 2,0 - 2,5 % et que le rapport K/Ca + Mg ne soit pas supérieur à 2,2. L'abaissement de l'absorption de magnésium par les plantes qui découle de l'application de potassium est bien connu. Les publications spécialisées mentionnent souvent deux niveaux critiques, à savoir la teneur en magnésium de l'herbe des pâturages, qui ne doit pas dépasser 0,2 % de la matière sèche, et la teneur en magnésium du sérum sanguin, qui ne peut pas être inférieure à 2 mg/100 ml. La teneur en magnésium de la matière sèche peut varier de 0,15 à 0,30 % selon la teneur protéique et potassique brute et le type d'aliments.

La tétanie d'herbage est certainement la plus connue des manifestations de carence en magnésium chez les ruminants, et est due souvent à la combinaison d'un abaissement de l'ingestion de magnésium avec les aliments et d'une diminution de l'absorption de magnésium par le système digestif. Les fortes laitières sont les plus exposées à la tétanie d'herbage. Cette maladie était

préoccupante dans les années 1950, mais depuis qu'on donne aux
vaches laitières des suppléments de magnésium et qu'on change
progressivement leur alimentation avant la mise au pâturage, cette
maladie est plutôt rare en Suède.

L'application d'engrais contenant 50 kg de magnésium à
l'hectare s'est traduite par une augmentation de la concentration de
magnésium dans l'herbe de 0,01 à 0,03 unité, alors que l'application
d'engrais potassiques (75-225 kg/ha) a fait diminuer la
concentration de magnésium de 0,01 à 0,05 unité (Baerug, 1977).

Indépendamment de l'effet fertilisant direct du potassium et du
magnésium sur les pâturages, d'autres facteurs ont une influence sur
l'équilibre en minéraux du fourrage, dont l'humidité du sol, la
composition botanique trèfle-herbe, le stade de développement et
l'application d'engrais azotés.

De façon générale, une humidité élevée du sol favorise
l'absorption du potassium au détriment du magnésium et du calcium,
d'où un coefficient accru K/Ca + Mg.

Pour ce qui est de la composition botanique, il est bien connu
que le trèfle a une teneur en magnésium beaucoup plus élevée que
l'herbe.

Le coefficient K/Ca + Mg est plus élevé au début qu'à la fin
des campagnes de récolte.

L'effet de l'application d'engrais azotés sur la composition
minérale du fourrage est d'autant plus complexe que l'ammonium (au
même titre que le potassium) entraîne l'abaissement de l'absorption
de magnésium par les plantes, alors que les nitrates favorisent
l'absorption d'éléments minéraux, en général. Un autre effet
largement connu de l'application d'engrais azotés est qu'elle
favorise la croissance de l'herbe au détriment du trèfle; l'effet de
ce phénomène sur la teneur en magnésium des herbages a été évoqué
plus haut.

SULPHUR REQUIREMENTS OF CROPS IN
THE UNITED KINGDOM

Mr. D. Hewgill
Ministry of Agriculture, Fisheries
and Food, Newcastle upon Tyne
United Kingdom

INTRODUCTION

Although sulphur has long been recognized as an important
constituent of plants, the United Kingdom has a relatively short
history of work on the sulphur requirements of crops. This is
because until recently very few cases of sulphur deficiency had
previously been diagnosed in agricultural crops. The reason for the
lack of sulphur deficiency in the past can be explained by:

(i)   Substantial depositions of sulphur resulting from the
       burning of fossil fuels;

(ii)   The use of NPK fertilizers which also contain sulphur;

(iii) Lower levels of crop yield.

The present interest in sulphur as a plant nutrient can be
traced back to 1956 when the Clean Air Act required that the levels
of sulphur dioxide emitted by industry should be reduced and that
taller chimney stacks should be used to disperse waste gases over a
wider area. Since this date there has been a steady decline in the
level of sulphur emission. Over the same time scale there has been
a sharp decline in the use of sulphur containing fertilizers such as
ammonium sulphate and single superphosphate and a corresponding
increase in the use of low sulphur fertilizers, for example ammonium
nitrate, urea and triple superphosphate. Average crop yields have
also increased considerably, thereby increasing the demand for
sulphur.

Numerous experiments and surveys were undertaken during the
1960s and 1970s to study the need for sulphur fertilizers in
relation to the declining level of sulphur pollution and increased
removal by crops (Bolton and Benzian 1970, Armitage 1972,
Williams 1975, Holmes 1977 and 1978, Whitehead et al. 1978). The
general conclusion of this work was that sulphur deficiency was
still unlikely in most areas of the United Kingdom.

As sulphur levels have continued to decline in the 1980s further studies have been carried out. In contrast to the earlier work, some of these experiments have shown yield responses to sulphur in rural areas of England, Wales and Scotland where sulphur pollution is low (Scott et al. 1983, Skinner 1985). This trend suggests that as sulphur dioxide pollution continues to decline the need for sulphur fertilizers is likely to become increasingly important, particularly in areas of low sulphur pollution.

This paper summarizes the factors affecting the sulphur nutrition of crops and gives a brief account of some of the more recent studies carried out in the United Kingdom.

CROP REQUIREMENTS

Plants normally take up sulphate-sulphur from the soil through their root systems. However on deficient soils some crops may also absorb up to half of their needs through their leaves as sulphur gases.

Sulphur is required by plants for protein synthesis. It is a constituent of the amino acids cystine, cysteine and methionine and is found in the primary enzyme RuBP carboxylase required for carbon dioxide assimilation. Certain crops species such as brassicas and onions also contain large quantities of sulphur containing metabolites such as glucosides and mercaptans and hence have higher sulphur demands than most other crops.

Sulphur deficiency symptoms in broadleaved plants are typically yellowing of the youngest leaves while the oldest leaves remain green. This differs from nitrogen deficiency which causes a similar chlorosis of the older leaves while the younger leaves remain green. With grasses and cereals sulphur deficiency results in a pale colouration in leaves of all ages and if the deficiency is severe the plants become weak and spindly. Legumes frequently develop reddish tinges on their leaves and nodulation is often retarded. Leaf tissue typically has a nitrogen : sulphur ratio of less than 15:1 and higher ratios usually indicate sulphur deficiency.

Sulphur deficiency can also delay harvesting and affect crop quality. Wheat flour with a N:S ratio of greater than 17:1 is likely to be deficient in cysteine and hence unsuitable for bread-making purposes. An acute shortage of sulphur can also reduce the nutritional value of crops by decreasing the content of sulphur containing amino acids. This is particularly important if grain legumes are the major source of protein.

Table 1 gives some examples of the amounts of sulphur removed by high yielding crops in the United Kingdom. From such values it is possible to calculate the removal of sulphur for a given level of crop yield. When crops are grown in areas where the level of atmospheric deposition of sulphur is less than the amount removed by the crops consideration should be given to the need for sulphur fertilizers.

Table 1

Sulphur Requirements of Crops

Crop	Harvested Part	Yield Level (T/ha)	Sulphur Removed (kg/ha)
Cereals	grain	8	14
	straw	5	8
Oilseed Rape	seed	3	25
	whole crop	12	57
Potatoes	tuber	40	20
Sugar Beet	roots	40	15
	tops	20	12
Kale	whole crop	10 */	42
Cabbage	whole crop	45	40
Grass	cut herbage	12 */	27
Lucerne	cut herbage	10 */	28

*/   Yield expressed on dry matter basis.

SOURCES OF SULPHUR

Virtually all of the sulphur utilized by crops comes from the soil, fertilizers and organic manures, or atmospheric sources.

Soil

The mineral fraction of soil usually contains little sulphur and that which is present generally has a low availability. Almost all of the plant available sulphur in soils is found in the organic fraction but probably less than 10 per cent of this is mineralized

during the growing season. This corresponds to as little as
3-5 kg/ha S per year for an intensive arable soil (Chaney and
Kershaw 1986) and around 8-18 kg/ha S for a grassland soil
(Keer et al 1986). Some sulhur originating from the mineralization
of organic matter or atmospheric deposition may also be present in
soils as sulphate-S. As free sulphate-S is about 60-75 per cent as
mobile as nitrate and chloride, any residues are easily leached from
soils during the winter period.

Inorganic fertilizers and manures

The last 30 years has seen the virtual disappearance of
straight N and P fertilizers which are rich in sulphur and the
emergence of types which have a very low sulphur content. However,
most NPK compound fertilizers still contain small amounts of sulphur
and typically contribute around 10 kg/ha S per annum at normal rates
of application. The sulphur contents of different types of
inorganic fertilizers used in the United Kingdom are given in
table 2.

Table 2

Sulphur Content of Inorganic Fertilizers at Percentage of Total Product Used

Fertilizer	Sulphur %	Total Usage %
Ammonium Sulphate	24	< 0.1
Ammonium Nitrate	< 1	42.6
Urea	< 1	0.5
Single Superphosphate	12	< 0.1
Triple Superphosphate	1	0.9
Potassium Sulphate	18	< 0.1
Potassium Chloride	< 1	0.4
NP, NK, PK, NPK	2-10	49.3
Others	–	6.3
	*/ Total Product ('000 tonnes) 5 471	

*/   Data from Survey of Fertilizer Practice 1985.

Organic manures can provide a further source of sulphur but information on their sulphur levels is limited. Values of 0.07, 0.08 and 0.22 per cent S are quoted for cow, pig and poultry manures respectively (Banwort 1975), while values of 0.07-0.10 per cent are given for cow slurry (Murphy 1983). Assuming that about 30 per cent of the sulphur is available to crops during the year of application it is estimated that 25 tonnes cow manure will supply crops with 5-6 kg S and 15,000 litres cow slurry with 3-5 kg S. If manures are applied during the autumn or winter months it is likely that much of the immediately available sulphur will be lost by leaching.

Atmospheric sulphur

Atmospheric sulphur originates from various man-made and natural sources. In the United Kingdom the major source is sulphur dioxide produced by the combustion of fossil fuel. Small contributions of sulphur also come from natural releases such as volcanic activity, while sea spray can contribute up to 15 kg/ha S in coastal areas.

Since 1970 the annual emission of sulphur has decreased by one third and it will probably decline further in the near future (Roberts and Fisher 1985). Of the 2 million tonnes of sulphur emitted in 1982, 65 per cent came from power stations, 28 per cent from other industry and 11 per cent from low level domestic sources (Buckley-Golder 1984). About 40 per cent of the total emission is deposited on crops or soil by dry deposition and 15 per cent by rainfall. The remainder is deposited beyond the United Kingdom coastline.

Dry deposition dominates in the industrial lowlands while wet deposition is more important in areas remote from major sources. A map of total wet plus dry deposition prepared by computer modelling is provided in figure 1. This shows deposition values ranging from over 40 kg/ha S in much of central England to below 20 kg/ha S in parts of South West England, Wales and Northern Scotland. Slightly less than half of the deposition falls in the summer six months.

Recent studies in Ireland (Murphy 1980) have shown that sulphur deficiency commonly occurs where the total deposition is below 15 kg/ha S per year. As sulphur deposition in the United Kingdom approaches these values it may be anticipated that responses to sulphur fertilizers will be obtained on some soil types.

PREDICTION OF SULPHUR DEFICIENCY

The following analytical methods are used to predict possible sulphur deficiency in areas where the level of sulphur deposition is less than crop requirements.

Leaf sulphur analysis

Plant sulphur levels vary with stage of growth, maturity, kind of tissue and levels of other nutrients. However if young leaves are sampled the difficulties of interpretation can be minimized. Suggested critical levels of sulphur in leaf dry matter (Skinner 1983) are given in table 3.

Table 3

Interpretation of Sulphur Levels in Leaf Dry Matter

Crop	Deficiency Likely (% S)	Deficiency Possible (% S)
Grass	<0.20	0.20-0.25
Clover	<0.20	0.20-0.26
Lucerne	<0.20	0.20-0.23
Wheat	<0.15	0.15-0.20
Brassicas	<0.27	0.27-0.30
Potatoes	<0.10	-

Plant N:S ratios

Critical N:S ratios in different plant tissues have been defined by many workers. In most cases ratios of more than 15:1 are considered to be indicative of sulphur deficiency. Others have found that critical N:S ratios can vary widely with crop age and nitrogen supply. Unpublished studies by MAFF have shown wide N:S ratios in crops growing in areas of high atmospheric pollution. No responses were obtained to sulphur fertilizers and it was concluded that the wide ratios were a reflection of high nitrogen usage rather than sulphur deficiency.

Water soluble sulphate

Plant water soluble sulphate values reflect the abundance of sulphur supply in relation to demand but have similar problems of interpretation as total S content. More than 500 mg/kg sulphate-S has been suggested for the optimum yield of grass (Scott et al. 1983).

Water soluble sulphate as percentage of total sulphur

This relationship is considered to be less affected by plant age and nitrogen supply. A critical value of 30 has been suggested for grass (Scott et al. 1983).

Soil analysis

It is generally agreed that phosphate extractable sulphate-S provides the best measure of plant available sulphur in soil. Potassium dihydrogen phosphate and mono calcium phosphate are used (Scott et al. 1983, Keer et al. 1986) as extractants and the critical value for both of these methods is 10 mg/kg sulphate-S. Calcium tetra hydrogen diorthophosphate has also been employed to measure the sulphur status of soils (Chaney and Kershaw 1986) and a scheme of interpretation for this method is given in table 4.

Table 4

Categories for Soil Sulphur Values

Sulphate-S (mg/kg)	Sulphur Status	Comments
<4	Very low	Yield responses to sulphur might be expected with most crops
4-7	Low	Small yield responses might be expected with brassica crops and 2nd or 3rd cut silage
8-11	Moderate	The sulphate available in the soil should be adequate to supply the needs of most crops
>11	High	

The above methods can however only measure the sulphate content of soils at the time of sampling and do not take into account subsequent leaching losses. Nor do they include longer term additions of sulphur from the atmosphere or contributions from the mineralization of organic matter. For these reasons some workers prefer to assess the sulphur status of soils on the basis of sand and organic matter content. Soils containing more than 50 per cent sand with less than 3 per cent organic matter are more likely to suffer from sulphur deficiency (Brogan and Murphy 1980).

CONTROL OF SULPHUR DEFICIENCY

Where sulphur deficiency is known to occur an application of 10-30 kg/ha S is recommended. This may be applied as 60-180 kg/ha gypsum ($CaSO_4$) or as finely divided elemental sulphur. Sulphur containing fertilizers such as ammonium sulphate (24 per cent S), single superphosphate (11 per cent S) and potassium sulphate (20 per cent S) may also be used where appropriate. Applications should be made at the start of the growing season since plants have a high demand early in their development. Autumn applications are not recommended because of the risk of leaching losses.

Where elemental sulphur is used on grassland it is recommended that the herbage should not be grazed until at least 10 cm of growth has occurred or 30 mm of rain has fallen. This is to minimize the risk of excess sulphur in the herbage inducing copper deficiency in livestock.

SUMMARY OF RECENT EXPERIMENTS AND SURVEYS

North of Scotland

The effect of sulphur fertilizer on the yield of herbage was investigated in 1980 and 1981 at two grassland sites with low available S levels. At rates of fertilizer N normally applied to well managed grass, sulphur additions increased dry matter production, particularly in the second and third cuts at high levels of nitrogen. Herbage total S, sulphate-S and N:S ratios confirmed sulphur deficiency in the absence of applied sulphur (Scott 1983).

Elemental sulphur, applied as a foliar spray to winter barley in March 1983, increased grain yields at four sulphur deficient sites by an average of 10.7 per cent. The increase was attributable to added S and not a fungicidal effect (Scott et al. 1984).

Sulphur deficiency in ryegrass was found to depress the concentration of the sulphur containing amino acids cysteine, and methionine as well as arginine, histidine, lysine, glycine, leucine, serine and threonene, thereby reducing the quality of crude protein (Millard et al. 1985).

East of Scotland

A survey of 147 farms in south-east Scotland in 1984 showed that around two thirds of the area had low sulphur soils. Sulphur demanding crops such as brassicas and conserved grass would be at risk from deficiency on these soils but less demanding crops such as cereals and potatoes are not at risk. In the Borders area 18 per cent of soils had a very low sulphur status and hence cereals and potatoes may also be at risk from sulphur deficiency. Soil texture and organic matter content were poor predictors of soils at risk from sulphur deficiency (Chaney and Kershaw 1986).

Sulphur experiments were carried out on winter barley, spring barley (eight sites), oilseed rape, swedes and grass (two sites) in 1984. There were no yield responses to sulphur fertilizers with cereals and oilseed rape on soils with low or very low sulphur contents. Significant yield increases were obtained with 80 kg/ha S applied to swedes and with 50 kg/ha S at one grass site cut three times (Chaney 1984).

West of Scotland

The effects of 0-48 kg/ha S applied as gypsum were examined at two grassland sites with low soil and herbage S levels in 1984 and four sites in 1985. No yield responses to sulphur were obtained in the first silage cuts in either year. Significant responses in dry matter yields were obtained in the second and third cuts at one site in 1984. There were also significant yield increases in the second cut at three sites and in the third cut at one site in 1985 (Harkness et al. 1986).

A survey of sulphur levels in soil and herbage samples carried out at 66 sites in 1983 showed that sulphur deficiency was a very minor problem. The results also indicated that deficiency problems were only likely to arise where sulphur demanding crops are grown on light free draining soils with low sulphur reserves in areas of low atmospheric sulphur inputs (Keer 1984).

England and Wales

The effects of 50 kg/ha S applied as gypsum in spring were tested at five sites with low atmospheric sulphur depositions in 1984 and at eight sites in 1985. Significant yield responses to sulphur were obtained in nine of the experiments. The pattern of responses was to find little or no yield response in the first cut but to obtain responses with subsequent cuts. Responses generally became larger as the season progressed. This suggested that sulphur accumulated during the winter months was rapidly depleted during the growing season (Skinner 1986).

Northern Ireland

Soil and herbage samples from 40 farms with coarse textured soils where nitrogen usage was high were taken in 1983. The extent of sulphur deficiency was assessed using water soluble sulphate values of 300–500 and 200–300 mg/kg S in herbage to indicate a yield depression of less than 5 and 10 per cent respectively. In first cut silage, 20 and 8 per cent of sites were predicted to have suffered yield depressions of less than 5 and 10 per cent respectively. Soil extractable sulphate values of less than 10 mg/kg S indicated marginal sulphate reserves for second cut silage at 49 per cent of sites. Analysis of second cut silage showed 11 and 3 per cent of sites were likely to have suffered yield depressions of less than 5 and 10 per cent respectively. As the farms were not a representative sample the assessment of sub-optimal sulphur is likely to be an over-estimate for the country as a whole (Stevens 1985).

CONCLUSIONS

There is now evidence of sulphur deficiency in parts of the United Kingdom where depositions of atmospheric sulphur are low. The incidence of sulphur deficiency is likely to increase further as the sulphur emissions continue to decrease. Further work will be needed to identify the crops at risk and to quantify the cycling of sulphur in different soil types and under varied farming systems.

44

REFERENCES

ARMITAGE E.R. (1972). Sulphur requirements of crops in the
United Kingdom. J. Sci. Fd. Agric., 23, p. 1150.

BANWORT W.L. and BREMNER J.M. (1975). Identification of sulphur
gases evolved from animal manures. J. Environ Qual., 4, pp. 363-366.

BOLTON J. and BENZIAN B. (1970). Sulphur as a nutrient for sitka
spruce seedlings and radish grown on a sandy podsol in England,
J. Agric. Sci. Camb., 74, pp. 501-504.

BROGAN J.C. and MURPHY M.D. (1980). Sulphur Nutrition in Ireland.
Sulphur in Agriculture, 4, pp. 2-6.

BUCKLEY-GOLDER D.H. (1984). Acidity in the Environment, ETSU,
Harwell.

CHANEY K. and KERSHAW C.D. (1986). The sulphur status of soils from
the south-east of Scotland. Research and Devel. in Agric., 3,
pp. 39-42.

CHANEY K. (1984). East of Scotland College of Agric. Annual Report,
p. 67.

HARKNESS R.D., KLESSA D.A., FRAME J. and GOLIGHTLY R.D. (1986).
Production response of grass swards to applied sulphur. Grass
Farmer No. 23, p. 20.

HOLMES M.R.J. and AINSLEY A.M. (1977). Fertiliser requirements of
spring oilseed rape. J. Sci. Fd. Agric., 28, pp. 301-311.

HOLMES M.R.J. (1978). Seedbed fertiliser requirements of winter
oilseed rape. J. Sci. Fd. Agric., 29, pp. 657-666.

KEER J.I. (1984). Is sulphur deficiency limiting crop yields in the
WSAC area? West of Scotland Agric. Coll. Tech. Note 228.

KEER J.I., McLAREN R.G. and SWIFT R.S. (1986). The sulphur status
of intensive grassland sites in southern Scotland. Grass and Forage
Sci., 41, pp. 183-190.

MILLARD P., SHARP G.S. and SCOTT N.M. (1985). The effect of sulphur
deficiency on the uptake and incorporation of nitrogen in ryegrass.
J. Agric. Sci. Camb., 105, pp. 501-504.

MURPHY M.D. (1983). Unpublished data.

MURPHY M.D. (1980). Much Irish grassland is deficient in sulphur. Farm and Food Research 11, pp. 190–192.

ROBERTS T.M. and FISHER B.E.A. (1985). Estimated trends in dry and wet sulphur deposition over the United Kingdom between 1970 and 1982 and the effects on crop growth. J. Sci. Fd. Agric., 36, p. 262.

SCOTT N.M., WATSON M.E., CALDWELL K.S. and INKSON R.H.E. (1983). Response of grassland to the application of sulphur at two sites in NE Scotland. J. Sci. Fd. Agric., 34, pp. 357–361.

SCOTT N.M., DYSON P.W., ROSS J. and SHARP G.S. (1984). The effect of sulphur on the yield and chemical composition of winter barley. J. Agric. Sci. Camb., pp. 699–702.

SKINNER R.J. (1986). Sulphur fertilisers for grassland – recent evidence. Grass Farmer No. 23, p. 23.

SKINNER R.J. (1983). Sulphur nutrition of crops in England and Wales. MAFF Report.

STEVENS R.J. (1985). Evaluation of the sulphur status of some grasses for silage in Northern Ireland. J. Agric. Sci. Camb., 105, pp. 581–585.

WHITEHEAD D.C., JONES L.H.P. and BARNES R.J. (1978). The influence of fertiliser N plus K on N, S and other mineral elements in perennial ryegrass at a range of sites. J. Sci. Fd. Agric., 29, pp. 1–11.

WILLIAMS C. (1975). The distribution of sulphur in the soils and herbage of N. Pembrokeshire. J. Agric. Sci. Camb., 84, pp. 445–452.

Figure 1

WET + DRY SULPHUR DEPOSITION

Kg/ha S per annum

1983 Data
Revised February 1986

LE 20.
LE 30.
LE 40.
GT 40.

(Map kindly provided by Dr B E A Fisher, Central Electricity Generating Boa
Leatherhead, Surrey.)

BESOINS DES CULTURES EN SOUFRE AU ROYAUME-UNI

M. D. Hewgill, Ministère de l'agriculture,
des pêches et de l'alimentation,
Newcastle upon Tyre

RESUME

On sait depuis longtemps que le soufre est un élément
constitutif important des plantes, mais les recherches entreprises
au Royaume-Uni sur les besoins des cultures en soufre sont
relativement récentes. En effet, jusqu'à ces derniers temps, un
petit nombre seulement de cas de carence avait été diagnostiqué dans
les plantes cultivées. Le fait que la teneur en soufre était
suffisante dans le passé peut s'expliquer pour les raisons
suivantes :

i)   La combustion des combustibles fossiles était à l'origine
      de dépôts substantiels de soufre.

ii)   Les engrais NPK qui étaient utilisés contenaient également
      du soufre.

iii) Les rendements agricoles étaient moins élevés.

L'intérêt suscité actuellement par le soufre en tant que
nutriment secondaire remonte à 1956, époque où la Loi sur la pureté
de l'air a exigé la réduction de la teneur en dioxyde de soufre des
fumées industrielles et la surélévation des cheminées d'usine afin
que les gaz expulsés soient dispersés sur une plus vaste étendue.
Depuis lors, le niveau des émissions de soufre a baissé
régulièrement. A la même époque, l'emploi d'engrais contenant du
soufre, comme le sulfate d'ammonium et le superphosphate simple
diminuait fortement, alors que l'emploi d'engrais à faible teneur en
soufre, comme le nitrate d'ammonium, et le superphosphate triple
progressait. De plus, les rendements moyens des cultures se sont
considérablement améliorés, ce qui a entraîné une augmentation de la
demande de soufre.

De nombreuses expériences et études ont été entreprises pendant les années 60 et les années 70 pour savoir s'il fallait utiliser des engrais contenant du soufre pour compenser la réduction de l'apport en soufre due au recul de la pollution et les pertes en soufre dues à l'accroissement des rendements. D'après les conclusions de ces recherches, les carences en soufre étaient encore peu probables dans la plupart des régions du Royaume-Uni.

Comme les teneurs en soufre ont continué à baisser pendant les années 80, de nouvelles études ont été entreprises. Contrairement aux précédentes, quelques expériences ont démontré qu'un apport de soufre améliorait les rendements dans les régions d'Angleterre, du Pays de Galles et de l'Ecosse où la pollution due au soufre était faible.

D'autres études ont montré que la teneur en soufre de certains sols et de certaines plantes était insuffisante. Comme la pollution due au soufre continue à reculer, il faudrait, semble-t-il, utiliser de plus en plus d'engrais contenant du soufre, notamment dans les régions où cette pollution est faible.

LIMING OF SOILS AS A FACTOR OF OPTIMIZING SOIL
CHARACTERISTICS AS WELL AS THE MAGNESIUM AND CALCIUM
CONTENT IN THE UPPER SOIL LAYER

Professor I.A. Shilnikov, Head of Laboratory, All-Union
Research Institute of Fertilizers and Soil Science,
Moscow

SUMMARY

The results of long-term field and pot experiments have shown
that losses of Ca and Mg from soil and soil acidity increased as a
result of agricultural intensification.

According to the results of pot experiments infiltrating waters
contained 65 to 75 per cent Ca and 18 to 24 per cent Mg (in total
cations) and more than 80 per cent SO_4^{-2}, CL^-, NO_3^- in
total anions. Average losses of Ca from the soil amounted
practically to 140-180 kg/hectare of Ca and to 20-30 kg/hectare of
Mg per year. This depends in many cases on various factors such as :
doses and type of fertilizers, characteristics of the soil, liming
of the soil, crop rotation, irrigation etc. The Ca content of the
soil has considerable influence on the frost resistance of plants.
Liming improves the resistance of plants to unfavourable weather
conditions. Doses of lime which are calculated very closely to full
hydrolytical acidity give the most efficient results.

Changes in soil pH depend on the doses of lime introduced into
soils with different physical structure. Introduction of one ton of
$CaCO_3$ changed soil pH as follows: on sandy soil 0.17-0.25, loamy
soil 0.12-0.19 and heavy loamy soil 0.09-0.11. The efficieny of
liming depends on the granulametric composition and on the solidity
of the ground lime. Utilization of dolomit meal is the most
efficient method to improve the magnesium available to plants.
Ashes from electrical power stations are a suitable material to
provide the soil with Ca and Mg. Granulated ashes from electrical
power stations are less effective in comparison with ground ashes.
Utilization of metallurgical slags, as lime fertilizers, meets the
requirements of plants of available forms of magnesium. Among
metallurgical slags, cast iron and electric steel slags have the
most positive influence on yields whereas Martin and especially
blast-furnace slags have a lesser effect on soil characteristics and
yields.

Table 1

Losses of Calcium in Connection with the Type of Nitrogen Fertilizer (kg/ha)

Years	Months	Without PK + lime	Variant of experiment		
			PK + lime		
			$CO(NH_2)_2$	$NaNO_3$	$(NH_4)_2S$
1	4-9	10.7	5.6	8.5	16.3
	10-3	99.5	78.4	73.7	120.2
2	4-9	49.4	49.0	55.2	147.2
	10-3	79.9	56.4	56.5	230.4
3	4-9	24.5	15.7	18.6	62.9
	10-3	26.8	23.3	22.3	95.0
4	4-9	17.0	11.7	15.5	73.2
	10-3	18.4	7.4	6.5	54.1
Average	4-9	25.4	20.5	24.4	74.9
	10-3	56.1	41.4	39.7	124.9

Table 2

The Influence of Lime Doses on Losses of Ca - Soil (kg/ha)

Years	Soil stratum in pots	Without lime	Dose of lime (meal) (g/kg)		
			0.5	1.0	2.0
1	0.35	48	49	57	59
	0.80	55	63	40	57
2	0.35	230	249	253	380
	0.80	201	215	125	177
3	0.35	88	95	97	106
	0.80	107	155	98	86
4	0.35	112	146	144	145
	0.80	89	102	70	67

Table 3

Average Annual Leaching of Ca and Mg by Drainage

Soil	Leaching by drainage (kg/hectare)	
	Ca	Mg
Derno – calcareaus heavy loam	0.95 – 1.71	0.24 – 0.65
Derno – podzolic light loam	0.86 – 1.32	0.24 – 0.41
Derno – podzolic sandy on loam	0.65 – 0.89	0.17 – 0.22
Peat – boggy well cultivated	1.20 – 1.40	up to 0.43

Table 4

Yield of Hay (1st year cut)
Acid and Limed Soils

Variant of experiment	Year of experiment							
	2nd	4th	8th	11th	2nd	4th	8th	11th
	Yield of hay				Including hay of clover			
	Quintal/hectare							
1. Without fertilizer	20.8	31.9	24.4	32.0	5.2	8.3	3.5	3.
2. NPK	17.4	30.4	22.2	32.0	0.5	1.8	0.0	1.
3. Lime	47.0	48.8	38.4	39.7	24.7	23.8	13.8	14.
4. NPK + lime	52.9	69.9	47.4	46.2	33.2	38.3	19.2	25.

52

Table 5

Average Annual Growth of Yield in Relation
to the Doses of Lime

Rotation	Doses of lime (rate of hydrolitic acidity)			
	0.25	0.5	1.0	2.0
First	$\frac{1.8}{37}$*	$\frac{3.5}{73}$	$\frac{4.8}{100}$	$\frac{6.2}{129}$
Second	$\frac{1.3}{23}$	$\frac{2.8}{50}$	$\frac{5.6}{100}$	$\frac{6.5}{116}$

* Numerator–feed units – quintal/hectare, denominator %.

Table 6

Efficiency and Quality of Lime Meal in Field
Experiments in the USSR

Average annual yield increase	Fractions of lime (mm)		
	Smaller 0.25	0.25–1.0	1.0–3.0
Feed units, quintal/hectare	16.6	14.8	9.4
Relative numbers	100	89	57

Table 7

Influence of Dolomit and Chalk on the Yield of Potatoes
and Lupins (average for 3 years)

Variant of experiment	Potatoes			Lupin	
	Tubers			Green mass	
	Yield	Increase		Yield	Increase
	Quintal/hectare		Starch %	Quintal/hectare	
1. $N_{60}P_{60}K_{60}$	276	–	15.3	436	
2. NPK + chalk	254	-22	15.6	405	-31
3. NPK + chalk + Mg_{50}	289	13	16.4	429	-7
4. NPK + dolomit	279	3	16.0	454	18
$HCP_{0.95}$,q/ha	18			14	

Table 8

Influence of Ashes and CaCO3 on Yields of Agricultural Crops

Background	Variant	Average of 26 yields		
		Yield	Increase	
		Feed units q/ha		%
NPK	Without lime	32.4	–	–
	Ashes	40.2	7.8	135
	$CaCO_3$	38.2	5.8	100
NPK + manure	Without lime	40.7	–	–
	Ashes	46.0	5.3	143
	$CaCO_3$	44.4	3.7	100

Table 9

Utilization of Granulated and Ground Lime in Liming

Type	Without lime	Increase of yield with	
		ground	granulated
	Feed units/ha	quintal/hectare	
Ashes of shale	36.7	6.6	4.8
Ashes of shale	36.7	6.6	1.9
Cement dust	32.6	10.7	3.6

Table 10

Effectiveness of Slags, Lime and Epsomit

Variants	Mgo applied	Winter rye	Oats	Grasses	Total yield for 6 years	Increase of yield from Mg	
	Kg/ha	Ca/ha			Feed units	q/ha	%
1. $N_{120}P_{180}K_{180}$	—	17.3	12.7	31.2	64.0	—	—
2. NPK + $CaCO_3$	—	20.3	15.7	45.7	76.6	—	100
3. NPK + slag NPK + lime	328	24.9	19.2	52.0	91.9	15.3	120
4. CaO + MgO — equivalent to slag	328	23.3	20.3	56.0	93.4	16.8	122
5. CaO + 1/5 MgO	66	24.7	18.5	57.0	93.9	17.3	123
6. CaO + 1/10 MgO	33	24.0	22.0	46.3	91.2	14.6	119
$HCP_{0.5}$		2.5	2.2	5.3			

PRESENCE AND DEFICIENCIES OF SECONDARY AND TRACE
ELEMENTS IN ARABLE AND GRASSLAND WITH HIGH
FERTILIZER USE AND VARYING SOIL CONDITIONS IN POLAND

Dr. H. Gembarzewski, Fertilizer and Soil
Research Institute, Pulawy

1. Introduction

Since the end of the Second World War, mineral fertilization
levels with NPK have increased remarkably. As a result crop yields
increased as well (Table 1). At the end of the 1980's fertilization
reached over 190 kg of $N + P_2O_5 + K_2O$ per ha of agricultural
area. In big state farms this level was much higher and in
1971-1972 it reached 253 kg/ha of agricultural area and 353 kg/ha of
arable land. In 1983-1984 it was as high as 281 kg/ha and 390 kg/ha
respectively. This indicates a considerable increase in crop
yields. In the case of wheat, the yield increased from 3.08 in
1971-72 to 4.25 t/ha in 1983-84 (Rocznik Statystyczny 1973 and 1985).

This development caused a large increase in the uptake of other
nutrients as well. Besides, the soil acidification by some mineral
fertilizer ions was further increased by SO_2 emissions from within
the country and from industrial sources localized abroad.

Part of the Polish soils - 67 per cent (Witek, 1981) -
constitute sandy soils appropriate for rye and potatoe cultivation,
with a small ion exchange and water capacity. In such conditions,
secondary and trace elements can become deficient.

In Poland deficiencies of secondary elements such as magnesium,
and - not as nutrient but as a soil amelioration factor - calcium
are expected. Sulfur deficiencies, mentioned by Katyal and Randhawa
(1983) are not generally expected in Poland.

The following trace elements: B, Cu, Mn, Mo, Zn are regarded
as important micronutrients for plant production in Poland. Because
of the small share of soils with a neutral or alcalic reaction (17
per cent), deficiencies of available Fe are very rare.

Data on elements necessary only for animals, considering their presence in plants and soils, have been scarce until now. This problem, however, could be solved successfully by oral application of the missing elements.

The soil contains relatively big total amounts of magnesium and micronutrients. Andruszczak's and Czuba's work (1984) informs about the total content of different elements in types and species of soils in Poland. Most of the sandy soils (with 0 - 20 per cent of particles sized mm < 0.02) contain 1 000 to 2 000 mg of Mg per 100 g of soil, which amounts to about 3 000 to 6 000 kg/ha of Mg in the 0 - 20 cm soil layer. Statistically the average yield of small grains in Poland amounts to 3.0 t/ha with a plant uptake of about 16.5 kg MgO (9.9 kg/Mg) per ha. The total Mg content estimated in the soil ploughing layer of 3 to 6 t Mg ought theoretically to suffice for obtaining 300 to 600 average yields of small grains.

According to Fotyma (1979) the average uptake of micronutrients by different crops in Poland is as follows:

Mn - 0.4 kg, Zn - 0.3 kg, B - 0.1 kg, Cu - 0.06 kg, Mo - 0.01 kg.

Considering that the average total content of these elements in the ploughing layer is for:

Mn - 900 kg, Zn - 150 kg, B - 60 kg, Cu - 45 kg, Mo 6 kg.

The following number of average crop yields could be produced theoretically:

Mn - 2250, Zn - 500, Cu - 750, B and Mo - 600.

Nevertheless, forms of these trace elements available for plants constitute only a small part of the total amount which means that plants often suffer from secondary nutrient deficiencies - especially for magnesium. Visual symptoms of periodical Mg scarcity on cultivated plants on sandy soils in Poland are common. In the case of micronutrients this is usually a "hidden hunger". This problem needs further research.

2. Soil fertility testing and current level of secondary and trace elements'utilization in field production in Poland

Together with the increased fertilization level in Poland, agro-chemical services were enlarged. At the beginning of the 1970s there were 17 Regional Agro-Chemical Stations to cover all of the country's needs. The stations are subordinated to the Institute of Soil Science and Plant Cultivation. They are testing soils on pH and P, K, Mg, B, Cu, Mn, Mo, Zn available for plants.

2.1 Soil reaction

Soil reaction effects immediately the plant growth and the availability of secondary and micronutrients. Figure 1 presents a map of acid soils in Poland according to provinces. In all the country 59 per cent of arable and grassland soils are acidic with a pH in 1 M KCl of up to 5.5, 24 per cent is slightly acidic and 17 per cent is neutral to alcalic.

Soil acidity has a fundamental significance for crop production, especially for improving the economy by way of enlarging the area under new, more soil acidity sensitive crops or cultivars, e.g. barley or alfalfa.

2.2 Soil fertility in regard to available magnesium and magnesium fertilization problems

At the end of the 1970's two cycles of Polish soil testing on the macronutrient content were completed. The available magnesium according to the Schachtschabel method was determined summarically in 14 753 000 soil samples. The results presented by Czuba (1979) were as follows:

Cycle	Percentage of samples with fertility level: (%)		
	low	medium	high
I. up to 1966	33	33	34
II. 1967 to 1977	44	34	22

The part of soils poor in magnesium increased during the same time in which, as a result of phosphorus and potassium fertilization, the part of soils poor in those macronutrients diminished from 56 to 47 and from 65 to 53 per cent, respectively. This is proof of the necessity of secondary element fertilization (in this case, with magnesium).

On the map (Figure 2) Mg deficiencies are shown according to provinces in Poland (After Czuba, 1979). They are concentrated in northern and central Poland.

In Poland different magnesium sources are used for fertilization. The author updated the data of Fotyma and Fotyma (1986) and estimated that Polish agriculture receives annually 83 600 t of MgO of which one third is in the dolomitized ammonium nitrate (with 4 per cent of MgO on a weight basis). Considering that as poor, and considering that soils medium fertile in Mg ought to be fertilized with magnesium, 78 per cent of the arable land and grassland must be subject to this treatment. The distribution of 83 600 t of MgO over the area of 14 771 000 ha results in 5.66 kg MgO per hectare. In this amount mixed fertilizers for gardens are not included. General recommendations for Mg application are 40 kg MgO and 20 kg MgO per ha per annum on soils poor and medium fertile in Mg, respectively. In such a case magnesium fertilizer supply is merely scarce.

It can be concluded that magnesium inadequacy is one of the important limiting factors in crop production in Poland, especially on sandy soils.

2.3 <u>Soil fertility in regard to micronutrients and micronutrient fertilizers supply</u>

Since the middle of the 1960's Agro-Chemical Stations have begun a regular survey on available micronutrients of arable and grassland soils. The following respective soil extraction methods have been applied: boron - with hot water according to Berger-Truog, copper - with 0.43 M HNO_3 according to Westerhoff, manganese - with the sulfite method of Schachtschabel, molybdenum - with oxalate according to Grigg, zinc - with 0.1 M HCl according to Sommer and Wear. The results were classified as "low", "medium", or "high". The appropriate threshold values were: Cu according to Westerhoff, B and Mn according to Bergmann, Mn according to Müller et al. (so called "Mo number" = pH in 1 M KCl + 10 x Mo in mg/kg), Zn according to Polish Soil Society. The results of the survey 1965 - 1983 were printed in the work "Zawartosc rozpuszczalnych mikroelementow w glebach Polski" (1985). One analysed soil sample represented about 33 to 46 ha. This is not sufficient for fertilizer recommendations, but it gives general information on micronutrient presence and deficiencies in a smaller administrative unit. As a whole, 400 - 570 thousand samples were analysed for all the 5 micronutrients: B, Cu, Mn, Mo, Zn. Generally, on a country scale the following results were received:

Element	Part of agricultural area with fertility level in		
	low	medium	high
boron	41	41	18
copper	41	25	34
manganese	25	14	61
molybdenum	44	49	7
zinc	9	39	52

Deficiencies of three elements: Mo, B, and Cu are significant. Mn availability depends mainly on the soil reaction and in the case of acid soils it is high or even excessive. On the contrary, Mo is much less available on such soils.

In the work cited the results are presented in tables and on maps - in per cent of low element content according to administrative units named "gmina" (or equal) of which in Poland there are about 2 160.

In this paper results of three micronutrient surveys are shown for only 49 provinces (Figures 3 - 5). A classification of the data according to administrative units was made in order to make future microfertilizer distribution easier. Apart from the great variety of microfertilizers for gardening purposes, the only micronutrient fertilizer produced on the industrial scale until today is triple superphosphate (TSP) with 0.5 per cent of boron.

Distribution of micronutrient-deficient areas is as follows: Boron - 11 provinces with over 50 per cent of soils poor in this element, localized in central eastern Poland. Taking into account that manufacturing of TSP with boron amounts to about 120 000 t per annum, 400 000 ha can be fertilized with a rate of 1.5 kg of B per ha. Assuming a recommendation per rotation at a rate of 1.5 kg B per ha under crops with high demands for boron, about 5 930 thousand ha ought to be fertilized (41 per cent of arable land poor in B). This is three times more than the area fertilized today. As a matter of fact, fields medium fertile with boron ought to be fertilized to some extent as well. In such a situation more economic ways of fertilizing, such as e.g. leaf application could be recommended. Copper - 21 provinces with 51 to 75 per cent of soils poor in Cu including one with more than 75 per cent are situated in north-eastern and central Poland. It is remarkable that 2 city provinces - Warsaw and Lodz - have a higher Cu fertility level than the provinces surrounding them. This is probably connected with industrial emissions of Cu. Molybdenum - 19 provinces in central Poland have 51 - 75 per cent of soils poor in Mo, including two over 75 per cent.

North-eastern Poland, where peat soils are common, is poor not only in Cu but also in Zn.

3. Evaluation of secondary and microelement deficiencies from plant chemical composition surveys and field trials

In 1976 the Agro-Chemical Stations (ACS) governed by their Methodical Centre in Wroclaw organized a network of so-called "control farms" (Czuba and Andruszczak, 1983). From two fields of each, 826 control farms soil samples (taken periodically) and crop samples (each year) are gathered and analysed according to separate cultivars. The fields are fertilized according to ISSCP recommendations. After summing up the data, Czuba and Andruszczak (1983) did not find any evident relationship between the NPK fertilization level and micronutrient content in plant material at harvest. This conclusion is drawn from comparing these data with the ones gathered by the ACS system 10 years earlier (Kaminska, Kardasz, Strahl, 1976). The authors of the cited work also published results of a survey on the chemical composition of hay (2 500 samples from all over the country). The hay samples contained in dry matter the following amounts of mineral elements:

Element and unit of content	Mean content	Desired minimal level for herbivorous animals
CaO – %	1.0	1.0
MgO – %	0.34	0.40
Mn – mg kg^{-1}	110	50
Cu – mg kg^{-1}	5.97	5.0
Mo – mg kg^{-1}	0.82	0.10
Zn – mg kg^{-1}	50	50
Fe – mg kg^{-1}	144	60
Co – mg kg^{-1}	0.08	0.05

According to the mentioned work hay in Poland is generally sufficient in micronutrients for animal requirements, except Cu, demands of which are actually higher (over 8 mg kg^{-1}).

According to the division of the country in 17 provinces which prevailed up to 1976, Cu deficiencies in hay were most common in Poznan (77 per cent of samles with < 5 mg/kg) and Bialystok (63 per cent of samples with < 5 mg/kg) provinces and cobalt in Bialystok (77 per cent of samples with < 0.05 mg/kg) and Szczecin (48 per cent of samples with < 0.05 mg/kg). The last two provinces are rich in meadows on peat soils.

Taking into account the situation in the whole country, 64 per cent of hay was poor in Ca, 67 per cent in Na, 70 per cent in Mg, 23 per cent in Mn, 28 per cent in Cu, 76 per cent in Zn. Besides some regional lack of Mg and Zn, the mentioned Cu and Co deficiencies seem to be a big problem.

Other trace elements necessary only for animals - selenium and iodine - were not surveyed. However, due to their easy leaching from soils, shortages could be expected.

The estimated needs of micronutrient fertilization are in agreement with the information from cooperating neighbouring countries (German Democratic Republic, Czechoslovakia). Boron fertilization is applied in practice, but copper and, even more, molybdenum are still a subject of research.

Faber and Filipiak (1984), basing themselves upon 1 270 sample analysis of wheat from control farms found that low Cu content in grain (< 2.5 mg/kg) was present in 72 per cent from light soils, 26 per cent from medium and 2 per cent from heavy soils.

Gembarzewski et al. (1984b) analysed the data from control farms concerning sugar beet (260 samples). They found that 40.2 per cent of roots contained too little Cu for high yielding (4 mg/kg Cu in dry matter according to Finck, 1979).

Similar analysis was undertaken concerning molybdenum (1984a). In about 20 per cent of the sugar beet samples Mo was deficient according to Finck. Similarly, 33 per cent of red clover samples were Mo deficient.

To evaluate Cu deficiency in small grains, the Institute of Soil Science and Cultivation of Plants conducted 150 field trials with wheat and oats and the after-effect on rye and barley. A preliminary summing up of the data shows that 26 per cent of the wheat and 21 per cent of the oat fields in Poland ought to be fertilized with Cu. The expected yield increase is not less than 10 per cent. The problem of the effect of control yield levels on copper demands by plants is not solved in these trials. The reason is the relatively low control yield level - usually less than 5 t of grain per ha.

A long-term trial on meadows was conducted by Czuba and Murzynski (1986) with high fertilizer rates of up to 600 kg N + 90 kg P_2O_5 + 240 kg K_2O for 15 years. It occurred that on loamy sandy soils the highest NPK rates lowered the Mg and Cu content in grass, and after more than 10 years it lowered the Zn content as well.

It ought to be mentioned that some macroelements
(macronutrients) influence the availability of some micronutrients,
such as N for Cu (Tisdale, Nelson and Beaton, 1985), while, for
example excessive P rates have a negative effect on Zn and Cu
(Gembarzewski et al., 1986) or a positive effect on Mo (Henkens,
1972).

4. Atmospheric pollution effect on the secondary and micronutrient
    balance

The effect of industrial pollution on soil fertility in copper
in two city provinces of Poland has already been mentioned. This
factor influenced the micronutrient balance.

In 1968 Chojnacki published data on the chemical composition of
precipitation in Poland. He found that even in agricultural regions
the upper soil layer "receives" annually 11.4 kg of Ca, 14.8 kg of
S, 2.0 kg of Mg, 3.8 kg of Na, 4.1 kg of Cl, 23.7 g of B per ha.

In 1980 – 1982 Kabata-Pendias et al. (1983) stated that in the
Pulawy surroundings a dustfall brings the following amounts of
various elements: 19–60 g of B, 21–32 g of Cu, 358–482 g of Zn,
138–1 699 g of Mn per ha.

Ruszkowska et al. (1984) conducted multi-annual lysimetric
research in the town of Pulawy. NPK rates recommended for
agricultural practice and lowered rates were applied on 4 different
soils. Commercially available fertilizers themselves contained some
amounts of micronutrients as admixtures. Potassium chloride and
ammonium nitrate were richest in Mn (up to 832 mg/kg) and Zn (up to
890 mg/kg). Superphosphate was richest in Cu (up to 27 mg/kg) and
Mo (up to 18 mg/kg). From rainfall there was 300 g of B, 225 g of
Cu, 1300 g of Zn, and 150 g of Mn per ha annually. However,
leaching of the elements was also remarkable. It ranged to:
7.5–295.0 g of B, 2.8–41.5 g of Cu, 0.5–72.5 g of Mn, 157–700 g of
Zn per hectare.

The supply from dustfall and rainfall and from macrofertilizer
admixtures balanced plant uptake and losses by leaching of B, Cu,
Mn, Zn. In the case of Mn and Zn, increasing NPK rates given in
commercial fertilizers improved the balance, while the same factor
spoiled it in the case of Bu, Cu and Mo. For Mo on light and loess
soil the balance was negative.

Results presented indicate the need for different treatments of
urbanized and unurbanized areas in the near future.

5. Conclusions

1. Great attention ought to be paid to micronutrient needs of the lands of central and north-eastern Poland, and still greater attention to light and organic soils all over the country.

2. The most urgent problem is to increase the supply of agriculture in Poland with lime, magnesium and boron and to begin the mass-scale manufacturing of copper and molybdenum fertilizers or to supply technical salts. The same urgency applies to the problem of animal diets' enrichment in the other necessary trace elements.

3. In the present situation, a more economic way of secondary and trace nutrients application ought to be applied in practice, namely leaf and seed application.

4. The micronutrient balance research ought to be developed on the territories with differentiated urbanization and industrialization levels.

64

Table 1

Fertilizer Consumption and Small Grain Yields in Relation to Agricultural Land in Poland in the course of Time

Year	Fertilizer use in kg ha^{-1}		Yields of grain in t ha^{-1}
	N+P$_2$O$_5$+K$_2$O	CaO	
1937/38	4.9	0.4	1.14
1949/50	17.7	6.2	1.23
1959/60	36.5	12.4	1.29
1969/70	123.6	90.9	2.16
1979/80	192.9	159.7	2.35
1983/84	182.5	154.33	3.00

Besides, in farmyard manure: 27 kg N + 20 kg P$_2$O$_5$ + 38 kg K$_2$O + 9.5 kg MgO + 25 kg CaO and 250 g Mn + 92 g Zn + 26 g B + 25 Cu + 3.4 g Mo + 1.1 g Co.
Per ha of arable land per year.

References

Andruszczak, E. and Czuba, R. (1984):

   Preliminary Characteristics of the Total Content of Macro- and
   Microelements in Polish Soils. (Roczn.glebozn. XXXV/2, 61-78.

Chojnacki, A. (1968):

   Results of Investigations on Chemical Composition of
   Atmospheric precipitations in Poland.   Part IV, Pam.pul. 35,
   163-172.

Czuba, R. and Andruszczak, E. (1983):

   Content of Microelements in Crops in the Country Network of
   Control Farms.   Zesz.probl. Post.N.roln. 242, 91-105.

Czuba, R. and Murzynski, J. (1986):

   Effect of 15 Years Differentiated Intensive NPK Fertilization
   on Meadow Yields and Chemical Composition in hay.   Part I-IV.
   Manuscript. ISSCP Wroclaw.

Finck, A. (1979):

   Dünger und Dünger, Verlag Chemie, Weinheim, New York.

Fotyma, M. (1979):

   Nawozy mineralne i nawozenie.   (In English: Mineral
   Fertilizers and Fertilizing).   PWRiL Warszawa.

Fotyma, E. and Fotyma, M. (1986):

   Wplyw terminu nawozenia na wielkosc plonow roslin.   (In
   English: Effect of Fertilizing Timing on Yielding Level).
   PWRiL Warszawa.

Faber, A. and Filipiak, K. (1983):

   Zawartosc mikroelementow w glebie w ujeciu przestrzennym.   (In
   English: Area Distribution of micronutrients in Soil).
   Manuscript.   ISSCP Pulawy.

Gembarzewski, H., Gromadzinski, A., Ferlas. A., Pelczar. M., Lipski, R. (1986):

   Phosphorus Effect on Copper and Zinc Assimilability for Plants. Zeszyty Naukowe Akad. Ekonom.we Wroclawiu (in print).

Gembarzewski, H., Stanislawska, E., Kaminska, W., Andruszczak, E., Kozlowska, H. 1984:

   Studies on the Availability of Molybdenum to Plants. Pam.pul. 82, 99-116.

Gembarzewski, H., Sowinska, B., Andruszczak, E., Kaminska, W. (1984):

   Availability of Copper for Sugar Beets and Winter Wheat in the Soils of Poland. Pam.pul.82, 117-130.

Henkens, Ch. (1972):

   Molybdenum Uptake by Beets in Dutch Soils. Centre for Agricultural Publications and Documents, Agric. Res. Rep. 775, Wageningen.

Kabata-Pendias, A., Tarlowski, P., Dudka, S. (1983):

   Opad pierwiastkow sladowych z atmosfery na powierzchnie gleb, Materialy z konf. "Racjonalne wykorzystanie gleb podstawa wyzwwienia narodu. PTGleb. Pulawy.

Nawozenie (1979). (In English: Fertilization) (Collective work under edition of R. Czuba). PWRiL Warszawa.

Rocznik Statystyczny (1973). GUS Warszawa.

Rocznik Statystyczny (1985). GUS Warszawa.

Ruszkowska, M., Rebowska, Z., Kapusta, A., Kusio, M., Sykut, S. (1984):

   Balance of Mineral Nutrients in a Lysimetric Experiment (1977 - 1981). III. Balance of Trace Elements (B, Mn, Cu, Zn, Mo). Pam.pul.82, 51-68.

Witek,T. (1981):

   Waloryzacja rolniczej przestrzeni produkcyjnej Polski, IUNG Pulawy.

Tisdale, S.L., Nelson, W.L., Beaton, J.D., (1985):

   Soil Fertility and Fertilizers, Collier Macmillan Publishers,
   New York.

Zawartosc rozpuszczalnych form mikroelementow w glebach Polski
(1985):

   (In English: Extractable Micronutrient Contents in the Soils
   of Poland).
   Editorial Board: Czuba, R., Gembarzewski, H., Debowski, M.,
   Klonowska, Z., Mleczko, E., IUNG Pulawy.

<u>Figure 1</u>

<u>Acid Soils in % of Agricultural Area in Poland</u>
(According to Provinces)

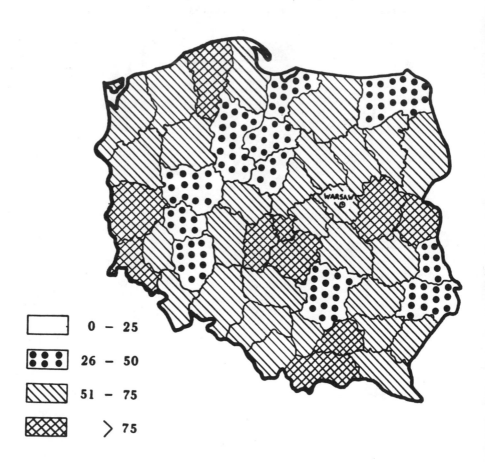

Figure 2

Soils Poor in Available Mg in % of
Agricultural Area in Poland
(According to Provinces)

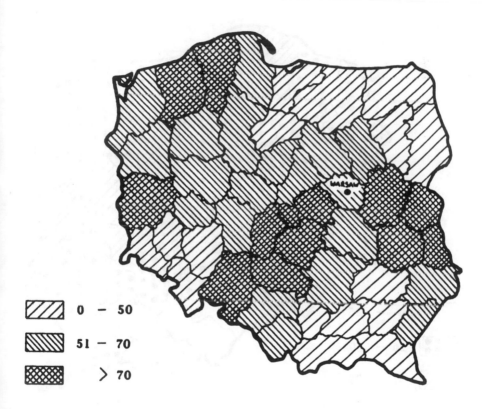

0 — 50

51 — 70

> 70

Figure 3

Soils Poor in Available B in % of Agricultural Area in Poland
(According to Provinces)

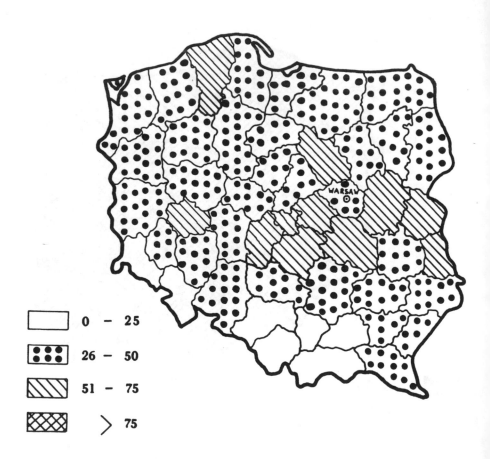

☐	0 - 25
⚅	26 - 50
▨	51 - 75
▧	> 75

Figure 4

Soils Poor in Available Cu in % of
Agricultural Area in Poland
(According to Provinces)

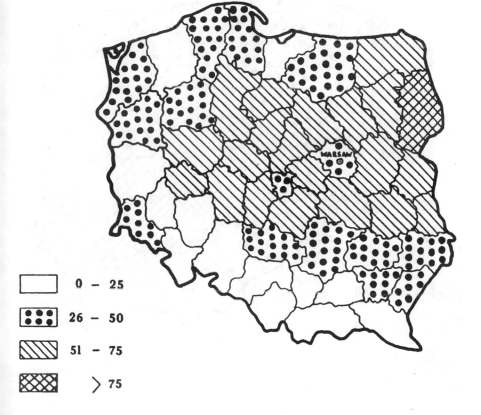

▢	0 – 25
⊡	26 – 50
▨	51 – 75
▦	> 75

Figure 5

Soils Poor in Available Mo in % of Agricultural Area in Poland
(According to Provinces)

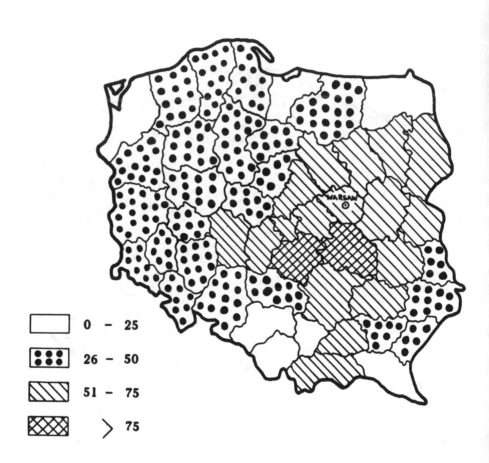

0 - 25

26 - 50

51 - 75

> 75

ELEMENTS NUTRITIFS SECONDAIRES ET OLIGO-ELEMENTS
PRESENTS OU DEFICITAIRES DANS LES TERRES LABOURABLES
ET PATURAGES A FORTE UTILISATION D'ENGRAIS
ET DANS DIVERSES CONDITIONS PEDOLOGIQUES EN POLOGNE

G. Gembarzewski de l'Institut de recherche
sur les engrais et les sols, Pulawy

RESUME

L'auteur présente les activités entreprises par les milieux
scientifiques et les services agrochimiques en Pologne pour évaluer
les effets d'une augmentation des doses d'engrais NPK sur
l'alimentation des plantes en autres macro- et oligo-éléments.
Depuis 1965, une double étude a été consacrée en Pologne au pH des
terres labourables et des pâturages, et aux apports de phosphate, de
potassium et de magnésium disponibles, ainsi qu'une étude
(préliminaire) sur la teneur du sol en bore, cuivre, manganèse,
molybdène et zinc. Il est prévu d'étendre le champ de ces
recherches aux régions pauvres en oligo-éléments. Par ailleurs,
l'Institut d'agrotechnique pour les engrais et la gestion des sols a
constitué en 1976, par l'intermédiaire de ses stations
agrochimiques, un réseau de 826 exploitations expérimentales.
Celles-ci étudient systématiquement les sols, le niveau des apports
d'engrais, les récoltes et leur composition chimique (teneur en
macro- et oligo-éléments).

D'après les résultats du deuxième cycle de recherches
répertoriées, 44 % des sols en Pologne sont pauvres en magnésium et
d'après 18 enquêtes entreprises pendant l'été sur les oligo-éléments
(500 000 échantillons analysés du point de vue de la teneur en
bore, cuivre, manganèse, molybdène et zinc disponibles), 41 % des
sols sont pauvres en bore, 41 % en cuivre, 44 % en molybdène,
25 % en manganèse et 9 % en zinc.

Les recherches effectuées dans notre pays sur le foin ont
révélé que 70 % du foin contiennent, du point de vue des besoins des
animaux, trop peu de magnésium, 76 % trop peu de zinc, 28 % trop peu
de cuivre et 23 % trop peu de manganèse.

Les données d'exploitations expérimentales montrent que 33 %
des échantillons de trèfle et 20 % de ceux de betterave à sucre sont
trop pauvres en molybdène pour une croissance satisfaisante. En
outre, 40 % environ des plantes-racines de la betterave à sucre
contiennent trop peu de cuivre pour assurer un rendement suffisant.
D'après les résultats d'essais de grande envergure sur le terrain,
20 à 26 % des céréales exigent un engrais contenant du cuivre,
surtout dans les sols légers. Le rapport est accompagné de cartes
schématiques sur l'acidification et les carences en magnésium, en
cuivre, en bore et en molybdène des sols polonais.

MICRONUTRIENTS AS RELATED TO THE SOIL
CHARACTERISTICS OF DIFFERENT COUNTRIES

Professor Dr. M. Sillanpää, Director,
Institute of Soil Science,
Agricultural Research Centre, Jokioinen

INTRODUCTION

Soil and plant samples from the most important agricultural
areas of thirty countries were collected for a micronutrient study
carried out in co-operation with the FAO and the Republic of
Finland. The material was analysed by one laboratory using the same
analytical procedure for all samples in order to obtain comparable
results.

Because of the global distribution of the sampling sites the
soil sample material represents growing conditions that vary greatly
and consequently, all soil properties are characterized by extremely
wide variations. This offers the unique possibility to study the
relationships between different soil characteristics and
micronutrients. Therefore in addition to micronutrient (B, Cu, Fe,
Mn, Mo and Zn) analyses, supplementary laboratory data were
collected from the samples (pH, electrical conductivity, organic
carbon content, texture, CEC, macronutrient contents, etc.) for
background information on the availability of micronutrients to
plants. The bulk of the results of the study have been presented
earlier (Sillanpää 1982). In this paper the typical properties of
soils in countries with a low micronutrient status are compared to
those having a high level of the respective micronutrients.

MATERIALS AND METHODS

The results of soil and plant analyses are often contradictory
as they are based on fundamentally different principles. Absorption
of micronutrients by plants is a process that takes place under the
laws of biochemistry and plant physiology, while chemical soil
extraction mainly follows the laws of chemistry. Therefore soil
analysis can be considered as an attempt to imitate plants, whereas
the amounts of nutrients measured from plants show what has really
been plant available in the soil. Consequently plant analysis is a
better basis for evaluating the role of soil factors affecting the
plant availability of micronutrients.

Soil and plant (wheat and maize) samples were collected from 3538 sites in different countries. The plant sample material, however, proved too heterogeneous to reflect reliably the micronutrient status of the soils where the plants had grown. To reduce the effects of this uncontrolled variation fresh samples of a new indicator plant (wheat, cv. "Apu") were grown in pots using the original soils. The plant data dealt within this paper are based on this plant material.

After drying, grinding, ashing and dissolving the plant ash Cu, Fe, Mn and Zn were measured by atomic absorption spectrophotometry. B content was determined by a modified azomethine-H method (Basson et al. 1969, John et al. 1975, Sippola and Erviö 1977) and Mo by the zinc dithio method (Stanton and Hardwick 1967).

Soil texture was determined by a pipette method (Elonen 1971) CEC by the Bascomb (1964) method, electrical conductivity from a soil : water (1 : 2,5) suspension, $pH(CaCl_2)$ from a suspension made 0,01 M with respect to $CaCl_2$, and organic C by modifying Alten's (1935) method (Tares and Sippola 1978). For more analytical details see FAO Soils Bull. 48 (Sillanpää 1982).

RESULTS AND DISCUSSION

Since micronutrients are not generally applied to the soil in the form of fertilizers their uptake by plants takes place mainly from their available fraction in soil. In special cases airborne micronutrients play a certain role. The amounts of micronutrients removed yearly with normal crop yields represent only a very small proportion, generally less than one per cent of the total amount present in soils. The total amounts, even in serious deficiency cases, therefore far exceed the crop requirements. Cases of primary deficiency that are mainly caused by a low total content of micronutrients, are therefore very rare in normal agricultural soils, but may occur in severely leached sands or in certain peat soils.

Secondary micronutrient deficiencies are the most common and are caused by soil factors reducing the availability to plants of otherwise ample supplies of micronutrients. For the effective correction in the field of a micronutrient deficiency, it is necessary to know which element is deficient and the reason why it is deficient. Therefore, a general background knowledge of soil characteristics is essential.

The effects of two or more different soil factors on the availability of certain micronutrients are often very similar (e.g. effects of electrical conductivity and pH on Cu, Table 1). In other cases the same soil factor affects the availability of two or more micronutrients very differently (e.g. effects of pH on Mo and Mn). Furthermore, it is often difficult to define which soil factor directly affects the availability of a certain micronutrient and which factor is only in "pseudocorrelation" with the micronutrient, owing to the mutual correlation between the two soil factors concerned. From the viewpoint of soil chemistry these questions are of importance, but in practical micronutrient studies a "pseudocorrelating" soil factor may often be as informative as a factor of direct effect.

In Table 1 three soil characteristics having a strong effect on the contents in plants of each of the six micronutrients studied are listed. Further, the values of these characteristics in the soils of two countries are given. One of these is a country where the plants have been able to absorb from soils only limited amounts of the micronutrient in question. The other is a country where the respective micronutrient exists in abundance.

Table 1

Some typical soil characteristics in countries of low vs. high micronutrients status

R = coeff. of correlation between the soil characteristic and resp. micronutrient content of plant in the whole material (n=3538); t = t-value of the difference between the two countries.

Micro-nutrient		R	Low	High (Micronutrient status)	t
Boron	Country		Nepal (n=35)	Iraq (n=150)	
	B cont. of plant (ppm)	.43xxx	4.1 ± 1.0	13.5 ± 14.9	7.66xxx
	pH(CaCl$_2$)	.59xxx	6.1 ± 0.9	7.7 ± 0.2	10.53xxx
	El. conductivity (10^{-4}S)	.19xxx	1.1 ± 0.5	9.4 ± 11.9	8.54xxx
	Org. C (%) cm		1.1 ± 0.5	0.8 ± 0.3	3.45xxx
Copper	Country		Sierra Leone (n=48)	Philippines (n=194)	
	Cu cont. of plant (ppm)	.36xxx	3.7 ± 1.2	10.5 ± 1.9	30.91xxx
	pH(CaCl$_2$)	.35xxx	4.9 ± 0.6	6.0 ± 0.8	10.38xxx
	El. conductivity (10^{-4}S)	.30xxx	0.8 ± 0.6	1.5 ± 1.2	5.64xxx
	Texture Index cm		33 ± 9	47 ± 18	7.64xxx
Iron	Country		Mexico (n=242)	Philippines (n=194)	
	Fe cont. of plant (ppm)	.14xxx	51 ± 13	79 ± 20	16.85xxx
	pH(CaCl$_2$)	.15xxx	7.2 ± 0.8	6.0 ± 0.8	15.19xxx
	Org. C (%)	.08xxx	0.9 ± 0.4	1.2 ± 0.7	5.08xxx
	CEC (me/100 g)		32.4 ± 14.4	35.5 ± 12.8	2.38xxx
Manganèse	Country		Egypt (n=198)	Zambia (n=44)	
	Mn cont. of plant (ppm)	.64xxx	41 ± 14	555 ± 451	7.56xxx
	pH(CaCl$_2$)	.41xxx	7.7 ± 0.1	5.0 ± 0.7	25.47xxx
	El. conductivity (10^{-4}S)	.27xxx	6.3 ± 4.9	0.8 ± 0.7	15.11xxx
	CEC (me/100 g) cm		43.2 ± 10.5	11.4 ± 6.3	26.32xxx
Molybdenum	Country		Brazil (n=158)	Pakistan (n=237)	
	B cont. of plant (ppm)	.46xxx	0.03 ± 0.04	0.68 ± 0.40	25.00xxx
	pH(CaCl$_2$)	.50xxx	5.1 ± 0.6	7.8 ± 0.3	56.25xxx
	El. conductivity (10^{-4}S)	.10xxx	0.9 ± 0.3	4.1 ± 6.4	6.67xxx
	Org. C (%) cm		2.0 ± 0.9	0.6 ± 0.3	19.18xxx
Zinc	Country		Iraq (n=150)	Belgium (n=36)	
	Zn cont. of plant (ppm)	.48xxx	11.1 ± 3.0	36.5 ± 21.4	7.10xxx
	pH(CaCl$_2$)	.29xxx	7.7 ± 0.2	5.7 ± 0.8	14.92xxx
	Org. C (%)		0.8 ± 0.3	1.2 ± 0.4	5.63xxx

The boron content of the test plant, wheat, correlates best
with electrical conductivity and pH of the soils[1]. These two
soil characteristics are also highly correlated with each other
($R = 0,578***$).

Soil pH has quite a small quantitative effect on the B content
of the plants within the acid and neutral pH range. A further rise
in pH, however, is followed by a substantial increase in plant B
content. Similarly the most marked rise in plant B occurs first at
a very high electrical conductivity level. A closer examination of
the international sample material showed that 84% of the 57 highest
el. conductivity values (exceeding 18 $10-4S/cm$) and 91% of the
highest plant B values (exceeding 15 ppm) originated from irrigated
sites, mainly in Iraq, Pakistan and Mexico. In Iraq the average B
contents of both plants and soils (hot water extractable)
originating from irrigated sites were 4 to 5 times as great as those
from the rainfed sites. Another factor correlating rather closely
with plant B is the organic carbon content of the soil[2].
B content decreases in soils with a high content of organic matter.
Quantitatively, however, this decrease is not substantial.

Obviously, B toxicity is most likely to exist in irrigated
fields with high electrical conductivity and pH, i.e. in conditions
that may be found e.g. in Iraq, Pakistan, Mexico, Syria and Turkey.
Although some low B values were measured from almost every country,
relatively these were most common in Nepal, Malawi, Nigeria,
Thailand and the Philippines.

---

[1] Also he B - $CaCO_3$ - equivalent correlation was high
($R = 0.338***$). Becuase of high mutual correlation between pH and
$CaCO_3$-eq. ($R = 0.903***$) and since over one third (n = 1228) of
the soil material (n = 3538) showed $CaCO_3$-eq. values approaching
0, the micronutrient/$CaCO_3$-eq. relationships are omitted.

[2] Because of the large sample material included in this study
(over 3500 pairs of samples), correlation coefficients as low as
0.033, 0.043 and 0.055 are considered significant at the 95, 99 and
99.9 per cent level, respectively.

The copper content of plants seems to depend on several soil factors. Among them, the best correlating are pH, electrical conductivity and texture[3/]. The role of pH as a factor affecting Cu solubility and/or its availability to plants is not clearly defined in the literature. This study, in contrast with many others, shows a steady though not drastic increase in plant Cu content from acid to the alkaline soils. A similar increasing tendency towards alkaline soils exists in the case of acid ammonium acetate-EDTA extractable soil Cu values. Since the soil total Cu contents were not measured, it cannot be defined whether the above relationships are due to respective increases in the total soil Cu or to increases in the relative availability of soil Cu.

Increases in electrical conductivity and texture index (toward the fine textured soils) values were also accompanied by increases in plant Cu content. According to the above relationships Cu deficiency would be most likely to occur in acid, coarse textured soils with low electrical conductivity. Such soils are common in all of the seven African countries studied and, in general, the lowest Cu values were measured from the sample materials originating in these countries, especially Sierra Leone, Zambia, Ghana and Ethiopia. High Cu values were measured especially in the materials from the Philippines, Italy and Brazil.

The iron contents of the test plants were quite insensitive to the six soil factors studied. Only a slight decrease for alkaline soil pH and slight increases with increasing organic matter content and CEC are to be mentioned. The correlations, although statistically highly significant, are weak compared to those in the cases of other micronutrients (Table 1). Contrary to the insensitivity of plant Fe, the amounts of chemically extracted soil Fe are strongly affected by the above soil factors and the respective correlation firmer (R-values 2-4-fold). However, as previously pointed out, chemical extractability and availability to plants are concepts which do not necessarily coincide. The oxidation-reduction conditions of soils are considered important in determining the behaviour of Fe in soils and its availability to plants. In this study the evaluation of these aspects was not possible.

3/ To express texture as a single figure a texture index (TI) was used. TI = 1.0 x % of fraction < 0.002 + 0.3 x % of fraction 0.002 - 0.06 + 0.1 x % of fraction 0.06 - 2.0 mm.

Although most of the lowest single plant and soil Fe values originated in Malta, Mexico and Turkey, indications of low Fe availability were also obtained occasionally from several other countries. Generally high Fe contents of plants were recorded from the Philippines.

When comparing the average soil properties of a typical low-Fe country, Mexico, to those of a high-Fe country, the Philippines, (Table 1) statistically significant but not drastic differences can be seen.

The manganese contents of plants decrease steadily with rising soil pH, electrical conductivity and CEC. The pH - plant Mn relationship is very close. The correlation is (R = 0.644***), in fact much higher than soil Mn - plant Mn correlations (correl. coefficients 0.552*** and 0.039* for DTPA and AAAc-EDTA extractions, respectively). Therefore, it is essential to take pH into account when estimating the Mn status of soils. On average, the Mn content of wheat grown on very acid soils is about ten times that of wheat grown on highly alkaline soils. Mn deficiency does not seem very likely on soils with pH($CaCl_2$) values lower than 6.0 - 6.5.

Correlations between plant Mn and the other two soil characteristics, electrical conductivity and CEC, are relatively high but clearly lower than that between Mn and pH.

Owing to the dominating role of pH in regulating the availability of Mn to plants, the most likely problems of Mn deficiency are to be found in areas with alkaline soils, while those of Mn toxicity are associated with acid soils.

Countries with a relatively high frequency of alkaline soils and a high probability of Mn deficiency include Malta, Egypt, Pakistan and Syria. Indications of Mn deficiency were rarely obtained from African countries, where acid soils predominate. In many of them, e.g. in Zambia and Malawi, an excess of Mn is a more likely problem as it might be in other such countries with acidic soils as Brazil and Sri Lanka.

The molybdenum contents of plants increased strongly toward the alkaline soils and contrary to Mn the Mo contents of plants grown on highly alkaline soils, were tenfold compared to those grown on very acid soils. This clearly indicates the important role of pH in regulating the availability of Mo to plants. The plant Mo - pH correlation is almost twice as strong as that between plant Mo and soil extractable Mo to ammonium oxalate-oxalic acid (r = 0.245***). Therefore, pH should be taken into account when interpreting the analytical results for available Mo.

Due to a high mutual correlation between pH and electrical conductivity, the relationship between plant Mo content and the latter soil factor resembles that between plant Mo and pH.

As in the case of boron, many high molybdenum values were obtained from samples originating from irrigated sites, often in Iraq and Pakistan. In Iraq the average Mo content of wheat grown on 75 irrigated sites was six times that from 44 rainfed sites.

In addition to the two countries mentioned, high Mo values were especially typical of Malta and Egypt. In all of these countries, alkaline soils predominate and the average $pH(CaCl_2)$ exceed 7.5. Soils with low Mo availability, low pH and low electrical conductivity are typical of Brazil, New Zealand, Nepal and most African countries.

Also in the case of zinc, pH plays an important role in regulating the availability of this micronutrient to plants although its effect seems not to be as drastic as in the cases of Mn and Mo. With rising $pH(CaCl_2)$ from about 4 – 5 to over 8, the Zn contents of potgrown wheat plants decreased steadily from about the 25 ppm level to the 10 ppm level, on the average.

There seems to be a tendency for plant Zn content to increase as organic carbon increases in the soils up to a level of about 6 – 8 per cent. With an increasing texture index, the Zn values first decrease, reach a minimum and then tend to increase again as the finest textured clay soils are approached. The shape of the curve may be the result of two opposing factors; increasing total Zn content and decreasing relative solubility in finer textured soils.

Although deficiency of Zn can be suspected somewhere in almost every country studied, it seems to be most widespread in Iraq, Turkey, India and Pakistan, but in several other countries the data indicate some shortages of Zn. High Zn values were especially typical of samples from Belgium, where industrial pollution may be partly responsible.

REFERENCES

Alten, F., Wandrowsky, B. & Knippenberg, E. 1935
Beitrag zur Humusbestimmung. Erg. Agric. Chem. 4: 61-69

Bascomb, C.L. 1964.
Rapid method for the determination of cation exchange capacity
of calcareous and non-calcareous soils. J. Sci. Fd. Agric. 15:
821-823.

Basson, W.D., Böhmer, R.G. & Stanton, D.A. 1969.
An automated procedure for the determination of boron in plant
tissue. Analyst 94: 1135-1141.

Elonen, P. 1971.
Particle size analysis of soil. Acta Agr. Fenn. 122: 1-122.

John, M.K., Chuah, H.H. & Neufeld, J.H. 1975.
Application of improved azomethine-H method to determination of
boron in soils and plants. Anal. lett. 8: 559-568.

Sillanpää, M. 1982.
Micronutrients and the nutrient status of soils: A global
study. FAO Soils Bull. 48. 444p.

Sippola. J. & Erviö. R. 1977.
Determination of boron in soils and plants by the azomethine-H
method. Finn. Chem. Lett. 1977: 138-140.

Stanton, R.E. & Hardwick, A.J. 1967.
The colorimetric determination of molybedenum in soils and
sediments by zinc dithiol. Analyst 92: 387-390.

Tares, T. & Sippola, J. 1978.
Changes in pH, in electrical conductivity and in the
extractable amounts of mineral elements in soil, and the
utilization and losses of the elements in some field
experiments. Acta Agric. Scand. Suppl. 20: 90-113.

LES OLIGO-ELEMENTS DANS LEURS RAPPORTS AVEC
LES CARACTERISTIQUES PEDOLOGIQUES DE
DIFFERENTS PAYS

Professeur Mikko Sillanpää, Directeur
de l'Institut de pédologie, Centre de
recherche agricole, Jokioinen

RESUME

   Des échantillons de sols et de plantes prélevés dans les zones
agricoles les plus importantes de 30 pays ont été analysés par un
laboratoire suivant les mêmes procédures afin d'obtenir des
résultats comparables.

   Comme les sites où ont été prélevés les échantillons sont
répartis sur tout le globe, ils représentent des modes de culture
très différents, et de très grandes variations dans les propriétés
des sols. Ces échantillons offrent la possibilité unique d'étudier
l'effet des oligo-éléments ("micronutriments") dans des sols ayant
des caractéristiques différentes. Aussi, en plus des analyses des
oligo-éléments (bore, cuivre, fer, manganèse, molybdène et zinc),
des données complémentaires ont été rassemblées en laboratoire sur
ces échantillons (pH, conductivité électrique, teneur en carbone
organique, texture, capacité d'échange de cations (CEC), teneur en
macro-éléments, etc.) pour obtenir des informations de base sur les
oligo-éléments dont disposent les plantes.

   Dans ce document, les propriétés caractéristiques des sols de
pays pauvres en oligo-éléments sont comparées à celles de sols
riches en ces mêmes éléments. Pour les six oligo-éléments
considérés, le pH du sol semble jouer un rôle essentiel dans la
régulation de leur disponibilité pour les plantes. C'est ainsi que
les pays dont les sols ont une faible teneur en bore, en cuivre et
en molybdène et une teneur élevée en fer, en manganèse et en zinc
ont des sols fortement ou modérément acides. La situation est
inversée dans les pays où les sols sont alcalins. Le rapport
existant entre la conductivité électrique des sols et la quantité
d'oligo-éléments présents est analogue à celui établi pour le pH,
car ces deux propriétés sont étroitement liées. Entre autres
facteurs liés à la disponibilité d'oligo-éléments figurent la teneur
en matières organiques, la capacité d'échange cationique et la
texture.

IMPROVED MANGANESE SUPPLY BY FERTILIZATION IN
HIGH-YIELDING CROPS IN THE NORTHERN REGIONS
OF THE FEDERAL REPUBLIC OF GERMANY

Prof. Dr. A. Finck,
Institute of Plant Nutrition and
Soil Science, Kiel
Federal Republic of Germany

INTRODUCTION

Micronutrient deficiencies (and toxicities) have been important
growth limiting factors for crops in the northern regions of the
Federal Republic of Germany since the last century, especially on
sandy or high-humic soils of low natural fertility. During the last
decades, however, most of the visible strong damage (e.g. of copper
deficiency) has been eliminated by soil fertilization.
Nevertheless, not only are there still numerous spots of
Mn deficiency in the sandy areas, but also in areas of high soil
fertility, after a yield increase to high levels (8-10 t/ha of
cereal grain), there is a larger and apparently ever increasing
latent and even acute Mn deficiency. It has actually become a major
limiting factor in high-yielding crops (after improvement of supply
with N, P, K, Mg) and is probably also responsible for lower
resistance of crops to environmental stress, e.g. cold damage.

Because of the complex nature of this Mn deficiency, it cannot
simply be cured by adding Mn fertilizers to the soil (as in the case
of copper). Manganese therefore presents a specific challenge to
agricultural chemistry.

EXTENT, CAUSE AND DIAGNOSIS

Mn deficiency symptoms are widespread in summer cereals (barley
and oats) and can easily be observed in June, mainly after a dry
spell of about two weeks (which is very common). In the
corresponding winter cereals the acute deficiency generally occurs
in less pronounced forms, whereas a latent deficiency seems to exist
on parts of many fields.

Other crops, too, are showing Mn deficiency symptoms,
e.g. maize, sugarbeet, oilseed rape. On grassland with a natural
carbonate content there may be just sufficient Mn for a growth of
plants without symptoms, but rarely sufficient for the higher
requirements of the grazing animals.

Figure 1 shows the occurrence of acute and latent Mn deficiency
of the soils of Schleswig-Holstein. For region B should be added
that the acute deficiency frequently occurs in relatively small
spots of several metres in diameter, but on many fields also larger
parts are more or less affected.

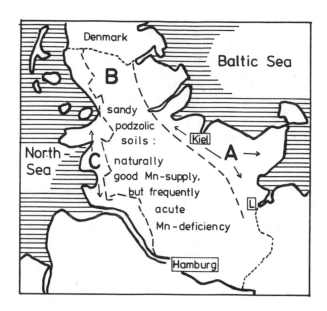

Legend: A Loamy Brown Earths ⎫ frequently latent
       C Alluvial Soils   ⎬ Mn-deficiency

Fig.1: Extent of Mn-Deficiency in Schleswig-Holstein

In humid climate with dominantly acid soils, for the medium to heavy textural soil types, the most important prerequisite for soil fertility improvement is proper liming, especially with respect to a stable soil structure. According to long-standing experience, optimum pH values have been established: pH 6 for loamy sands, pH 6.5 for sandy loams and pH 7 for loams and loamy clays (pH measured in 0.01 M $CaCl_2$, the data being about 0.5 pH lower than obtained in water suspension).

Whereas at these optimum pH levels the Mn supply was generally still sufficient at medium yield levels, it often becomes insufficient at high yield levels of 8-10 t/ha of grain (the minimum plant requirement being about 800 g Mn/ha). The decrease of Mn availability with increasing pH, as measured by exchangeable, easily reducible or complex extractable fractions, is very pronounced but differs according to soil properties (Fig. 2).

On formerly acid soils the widespread Mn deficiency apparently results from unintentional overliming (mainly due to unequal distribution) as well as from proper liming up to the "optimum" pH. It thus appears to be a "man-made" deficiency. On the other hand, in areas with neutral or calcareous soils (e.g. young alluvial), the Mn deficiency also seems to be of growing importance with increasing yields.

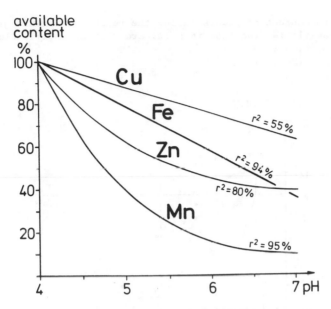

Fig. 2: Available micronutrients (DTPA-extract)
in a Brown Earth depending on soil
reaction (pot experiment)

SCHNUG, 1982

The diagnosis of Mn availability in soils has not been
established satisfactorily as yet. Soil testing in this respect
appears to be less reliable than for potassium of magnesium, and a
comparison of old and new methods is under investigation (Table 1).

Table 1

Correlation Between Available Mn and Plant Mn (as
calibration index) resp, pH (132 samples of oat
and wheat, preliminary data)

FLÜH, 1986

Method of soil extraction	B-value (r^2 in per cent) between available Mn and	
	Plant-Mn	pH ($CaCl_2$)
NaCl	47	54
NH_4OAc pH 7.0 */	46	35
$CaCl_2$	42	58
NH_4OAc pH 7	39	38
EDTA pH 8.6	35	35
DTPA pH 7.3	22	27

*/ field-fresh samples.

The Mn content of plants during the period of maximum growth, like the available fractions in soil, are also strongly influenced by soil reaction (Fig. 3).

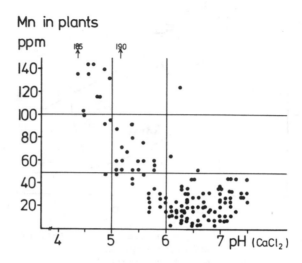

Fig. 3: Mn contents of oat and wheat depending on pH
(ppm in dry matter of above-ground parts at
beginning of shooting)

FLÜH, 1985

Plant analysis, on the contrary, provides a good index of the Mn supply (based on total Mn), e.g. for cereals the critical level is about 35-40 ppm Mn in dry matter at the first halm knot stage.

The actual nutrient supply of some fields is shown in Figure 4. On field A, in spite of an excessive fertilization with major nutrients, only medium yields are obtained (varying from 4-6 t/ha) due to Mn deficiency during many years whereas on field B with a good Mn supply maximum yields are possible (and often obtained).

Several colour photographs (which cannot be reproduced here) show the distribution of the deficiency in the fields, some acute as well as pre-acute symptoms in crops. The latter are the first and somewhat untypical changes of the normal green leaf colour in certain parts of the leaves. Furthermore, some "healthy looking" green fields are shown which are actually fields with latent deficiency (where fertilization with manganese would give a substantial yield increase).

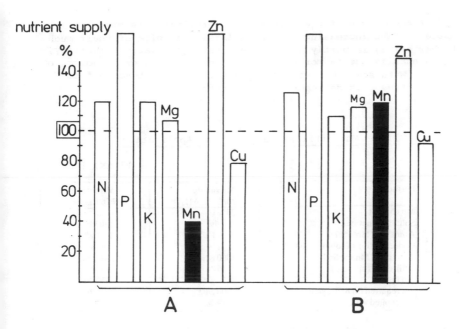

Fig. 4: Nutrient composition of young wheat plant in two
         farmers fields near Kiel (nutrient contents
         relative to critical limit, field A = medium
         yield, field B = high yield)

                                           FINCK, 1982

IMPROVEMENT OF THE MN SUPPLY BY FERTILIZATION AND MOBILIZATION

     Unlike copper deficiency where soil application of
Cu fertilizers almost always leads to improved plant supply in the
long-term, this is unfortunately not the case at once with manganese
on the problem soils. For the correction of Mn deficiency, however,
there are different possibilities with their advantages and
drawbacks.

(a) Leaf application of soluble Mn fertilizers

     Supplementary supply of the plants by leaf spray is common
practice in high-yielding crops. This method is effective if
applied in time and if repeated several times.

     This treatment is efficient, however, only for one vegetation
period and has to be repeated every year.

A comparison of the main single fertilizers is given in Table 2. The increasing use of chelates (in spite of their much higher price) is partly due to their very good water solubility, which enables the farmer to mix it quickly into a great volume of water. With some well waters, however, chelates tend to form precipitates thus decreasing their effectiveness.

Table 2

Mn Fertilization for Leaf Application

FINCK, 1984

Formula	Mn sulfate MnSO$_4 \cdot 4$H$_2$O	Mn chelate Mn EDTA
Mn content	24%	12%
Solubility in water	slowly	quickly
Amount of fertilizer	4 kg/ha	1.5 kg/ha
Concentration of solution (400 l/ha)	1 %	0.4 %
Amount of Mn applied	1000 g/ha	180 g/ha
Utilization (approx.)	15 %	35 %
Mn uptake	150 g/ha	63 g/ha
Mn required for 8 t/ha yield	800 g/ha	800 g/ha
Mn applied with 1 spray	19 %	8 %

The use of single Mn fertilizers requires a preceding reliable diagnosis (optical or chemical) which is demonstrated in Figure 5. In this case it would be useless to apply any other micronutrient except Mn.

Many farmers, however, have no exact knowledge of the micronutrient status of their crops and therefore tend to use combined fertilizers with all important trace elements, as well as the often yield limiting magnesium. This method is considered to be a certain "insurance policy", but cannot be as effective as the use of single fertilizers in case of a definite Mn deficiency. With 1 kg/ha e.g. of Fetrilon Combi (containing now 4 per cent of Mn) only 60 g of Mn are applied - compared with the amounts given in table 1.

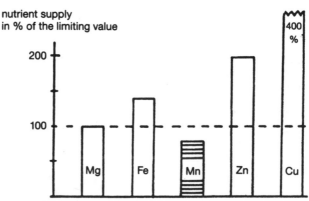

Fig. 5: Nutrient supply of wheat at the beginning of shooting (farmer's field near Kiel)

FINCK, 1984

The combined application of micronutrients will be useful in cases of very light deficiencies, but hardly sufficient even for medium deficiencies (unless repeated many times).

(b)   Soil application of Mn fertilizers

The effect of Mn fertilizers is generally significant on soils without deficiency, i.e. showing no strong immobilization (fixation) of water-soluble Mn compounds resp. on soils causing a distinct mobilization of water-insoluble Mn fertilizers, e.g. Mn oxides.

Unfortunately, on soils with medium or strong Mn deficiency (where fertilization would be definitely required) the fertilization effect of soluble Mn compounds is at best rather small, but mostly insignificant (water-insoluble fertilizers showing more or less no effect).

The addition of Mn fertilizers mainly increases the mobilizable Mn reserves in the soil which are, however, rarely in a limiting factor even on sandy soils.

On the other hand, since on calcareous soils a mobilization of soil Mn reserves by acidification (as will be described in the next chapter) is not very effective due to their high buffer capacity, Mn fertilization offers the only way of soil improvement. It seems to be worthwhile, therefore, to find an effective and lasting method of Mn fertilization.

Theoretically, the following approaches seem to be promising for an improvement of Mn fertilization on strongly deficient soils with poor Mn supply from the soil:

- placement of Mn fertilizer

- combination of Mn fertilizer with phosphate, both being placed together in the main rooting zone

- combination of Mn fertilization with acidifying N fertilization.

In Table 3 the data of a pot experiment are presented showing the yield effect and the better Mn uptake of placed fertilizers.

Table 3

Effect of Different Application Methods of
Water-insoluble Mn, Fertilizers on Growth and
Mn Content of Oats

RASMUSEN, 1981

No.	fertiliz. with N and P	A. Dry matter of young plants (g/pot)		
		Mn fertilization		
		no (control)	Mn-distrib.	Mn placed
1	M_0P_1	2.0	1.8	2.4
2	N_1P_0	3.1	3.4	4.3
3	N_1P_1	3.4	3.6	5.9
4	N_1P_2	3.7	3.8	5.6
5	N_2P_1	5.6	5.6	6.0
		B. Mn content of young plants (ppm of dry matter)		
1	N_0P_1	12	17	29
2	N_1P_0	7	9	16
3	N_1P_1	7	7	67
4	N_1P_2	8	9	51
5	N_2P_1	15	16	76

Legend: Calcareous humic soil, pH 7.4, 1 kg. pots
N = NH_4NO_3, P = Triplephosphate,
Mn – Mn-oxides-fertilizer plants = oats, cut at about
25 cm height; N_2 = standard dose

Neither the weight of the young plants (as indicator of growth)
nor their Mn content has been increased by well-distributed
Mn fertilizer of the non-watersoluble type mixed into the root
volume of the N fertilized pots.

The placement of Mn, however, together with phosphates
significantly increases the Mn content 2 to more than 4-times and
far above the critical Mn content (i.e. higher than 35 ppm). This
growth effect obtained is shown in a colour photograph.

These preliminary results of pot experiments have to be
confirmed by field trials. Even if the effects were less
pronounced, there seems to be nevertheless a possibility of
increasing the effect of Mn fertilization.

(c) Slight acidification of soil during the main growth period

The close relation between soil reaction and Mn uptake suggests the utilization of this effect for better soil supply with Mn. Actually, this method is already used by farmers without realizing it, namely by applying soil acidifying nitrogen fertilizers. Urea is more effective in this respect than nitrochalk, but ammonium sulphate is still superior in mobilizing Mn compounds. The effect of this procedure can be shown in pot and field experiments, only on non-calcareous soils however (Fig. 6).

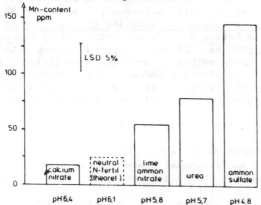

Fig. 6: The increase of Mn contents of young oat plants due to soil acidification by N fertilizers (pot experiment with Brown Earth; 0.4 g N/kg soil)

SCHNUG and FINCK, 1982

In some cases, a light deficiency can even be cured and some yield responses obtained with N fertilizers are mainly due to better Mn supply (Fig. 7).

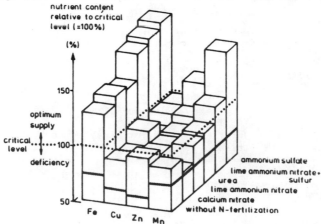

Fig. 7: Influence of different N fertilizers on micronutrients of summer wheat relative to the critical concentration (stage "ear appearance")

SCHNUG and FINCK, 1982

(d) Cropping at a lower optimum pH

The presently valid optimum pH data were established some decades ago when wheat yields were in the range of 4-6 t/ha of grain and when micronutrients were no limiting factor on most of the better arable soils.

Obviously, most medium to heavy soils, if limed only up to about pH 6-6.3, would hardly have any Mn supply deficit. However, the problem of structural stability at this level is not yet solved (though under investigation). A reasonable procedure seems to be the application of a mixture of lime and gypsum (instead of lime alone) in order to have a high Ca concentration in the soil solution without a strong pH rise. If such a procedure were practical, it would provide a simple method of avoiding Mn deficiency in large areas (as well as for zinc and boron). Furthermore, it would not be costly because of the large amount of gypsum now available from the desulphurication process at the electric power stations.

ECOLOGICAL ASPECTS OF Mn FERTILIZATION

The consequences of widespread Mn deficiencies in crops are not only yield losses and financial losses due to unused high input of other fertilizers etc., but also unwanted ecological effects.

If, for example, a wheat yield is fertilized (and provided with plant protection) in the expectation of high yields but produces actually only a 10 to 20 per cent lower yield, a part of the major nutrient fertilizers is added unnecessarily. It may either be inactivated or even be lost - thus Mn deficiency being responsible e.g. for an unnecessary water pollution by nitrate.

It therefore seems justified from several aspects to eliminate the yield-limiting factor Mn deficiency.

PUBLICATIONS
(research work and reviews on which manuscript is based)

Finck, A.: Besser und billiger düngen zur Ertragssteigerung und
Kostensenkung. Schriftenr. Agrarwiss. Fak. Kiel. 63, 39-51, 1982.

Finck, A.: Fertilizers and Fertilization. Introduction and
practical Guide to Crop Fertilization. Verlag Chemie Weinheim -
Deerfield Beach, Florida 1982.

Finck, A.: Micronutrient problems in high-yielding cropping.
Proc. 9. World Fertil. Congr. (CIEC, Budapest) 1, 263-267, 1984.

Flüh, M.: Diagnose des Mangan-Mangels (unpublizierte Daten) 1985/86.

Rasmusen, B.: Untersuchungen zur Verbesserung der Mangan-Düngung
(unpublizierte Daten). 1981.

Schnug, E.: Untersuchungen zum Einfluss bodenversauernder Düngung
auf die Spurennährstoff-Versorgung von Kulturpflanzen. Dissertat.
Agrarwiss. Fak., Kiel 1982.

Schnug, E., and Finck, A.: Trace element mobilization by acidifying
fertilizers. Proc. 9. Intern. Plant Nutr. Coll. 2, 582-587, 1982.

AMELIORATION, GRACE A LA FERTILISATION, DE L'APPORT
DE MANGANESE DANS LES CULTURES A GRAND RENDEMENT
DES REGIONS SEPTENTRIONALES DE LA
REPUBLIQUE FEDERALE D'ALLEMAGNE

Professeur A. Finck, Institut de recherche
en matière de nutrition des plantes et
de science du sol, Kiel

RESUME

Parmi tous les oligo-éléments, le manganèse est celui dont la
carence est devenue le principal facteur limitatif du rendement des
cultures arables. Ce problème est dû essentiellement au chaulage
des sols naturellement acides qui est pratiqué couramment par les
agriculteurs. On ne peut pas le résoudre durablement par une simple
adjonction de manganèse aux engrais (contrairement à ce qui se passe
pour le cuivre). Les céréales d'été et d'hiver et aussi d'autres
cultures souffrent de carences (aiguës ou latentes) de manganèse,
ce qui non seulement entraîne des pertes de rendement, mais
contribue aussi à réduire l'efficacité d'autres éléments nutritifs
fertilisants et produit des effets indésirables sur l'environnement.

Jusqu'à présent, le diagnostic de carence en Mn par l'analyse
de la teneur en Mn du sol était assez incertain. En outre, il était
difficile de reconnaître les symptômes avant-coureurs de cette
carence, avant qu'elle ne devienne aiguë. L'analyse des plantes
permet toutefois un diagnostic sûr puisque la teneur totale en Mn
des jeunes plantes reflète la teneur du sol en cet élément.

Les méthodes utilisées pour pallier les carences en Mn ne sont
pas encore satisfaisantes. L'application sur les feuilles de
composés solubles de Mn est efficace (si elle est suffisante et
fréquente), mais ses effets ne durent qu'une saison. L'apport
d'engrais manganesé dans les sols pauvres en manganèse n'est pas
très efficace et n'a pas d'effet durable. Toutefois, il semble
possible d'accroître l'utilisation des engrais manganesés à l'aide
de méthodes d'application spécifiques.

L'acidification des sols à l'aide d'engrais azotés s'avère très
efficace pour remédier à des carences légères sur des sols calcaires
trop chaulés. Enfin, pour maintenir un apport naturel suffisant de
Mn dans les champs à fort rendement, peut-être conviendrait-il de
cultiver les sols à un niveau de réaction inférieur au niveau
"optimum" comme c'est le cas actuellement, en corrigeant l'effet
structurel nécessaire par l'adjonction de gypse combiné à de la
chaux (en moindre quantité).

En résumé, l'amélioration de l'apport en manganèse par les engrais ou la correction du pH est un problème important, mais qui jusqu'à présent n'a été qu'imparfaitement résolu, dans les régions septentrionales de la République fédérale d'Allemagne où l'on pratique la culture intensive. L'effet des engrais manganesés sur les sols qui en sont dépourvus est insignifiant ou bien il n'est pas durable. Une solution prometteuse de ce problème semble toutefois passer par l'élaboration d'un concept nouveau de culture des sols à un niveau de réaction optimum (qui serait inférieur), c'est-à-dire à l'aide de méthodes d'application spécifiques des engrais.

INITIAL AND RESIDUAL EFFECTS OF
MICRONUTRIENTS IN CROP ROTATION

Dr. A. Faber, Institute of Soil Science
and Crop Cultivation, Pulawy

INTRODUCTION

The study which was aimed at determining the micronutrient
status of soils contained small or inadequate amounts of
micronutrients (Gembarzewski 1986). According to data provided by
the Agro-Chemical Stations soils with low Mo, Cu, B, Mn, and Zn
contents cover about 44%, 41%, 25%, and 9% of the arable lands,
respectively (Gembarzewski 1986). However, in the majority of
experiments carried out in Poland on a correction of the
micronutrient status of soils (Cu, Mn, Mo, and Zn) there was no
evidence of significant effects on yields. It is likely that
subclinical deficiencies of micronutrients occur under Polish
conditions. Of all the micronutrients boron alone is of practical
use in Polish agriculture.

It is difficult to identify a subclinical micronutrient
deficiency. The problem may be approached with greater precision by
way of soil and tissue testing. However, prophylactic fertilization
is the only practical way available at present. This fertilization
model is also determined in relation to the requirements and
tolerances of a crop, the properties of soils, the content of
micronutrients and the residual effectiveness of fertilization
(Bergmann 1976).

In the last years publication on the correction of
micronutrient deficiency have been numerous, but the residual
effectiveness of micronutrient fertilizers under field conditions
and the timing of reapplication of fertilizers were measured only in
a limited number of experiments (Brennan et al. 1986 a, b; Barrow et
al. 1985, Singh and Abrol 1985, Gupta and Cutcliffe 1984, 1982).
The field experiments reported here examined the initial and
residual effectiveness of micronutrients in crop rotation and the
initial effectiveness of micronutrients for two crop rotations in an
area with subclinical micronutrient deficiency.

MATERIAL AND METHODS

Randomized block experiments, comprising four levels of boron, copper, molybdenum and zinc, four replicates and four crops were conducted over six years (1979-1985). Micronutrients were applied to highly responsive plants in the crop rotation: sugar beet (B) – maize (Zn) – peas (Mo) – winter wheat (Cu). In the three subsequent years the residual value of the initial application for each micronutrient and for all succeeding crops were examined. The characteristics of the experimental field are shown in Table 1.

Micronutrients were applied as B – borax (11%), Cu – copper sulfate (25%), Mo – ammonium molybdate (54%) and Zn – zinc sulfate (23%) in solution two weeks prior to seeding at the following rates: (kg/ha):

boron	0,	1,	2,	4
copper	0,	5,	10,	20
molybdenum	0,	0.5	1.0	2.0
zinc	0,	5,	10,	20

They were incorporated to a depth of about 10cm. After the fourth crop further micronutrients were applied on the same plots and at the same rates in order to compare the initial effectiveness of the previous application (first rotation) with a current application (second rotation). The ratio and amounts of N, P, K, Ca, Mg fertilizer were consistent with those recommended by the National Computer System for Fertilizer Recommendations.

Plants were examined throughout the growing season for micronutrient deficiency and for toxicity symptoms. The crops were harvested when fully developed or mature. Marketable yields were recorded.

The plant samples were collected at the following stages: beet – 50 to 60 days after sprouting (central leaves), maize when in blossom (cob leaf), peas – commencement of flowering (whole plants), wheat – first node detectable (whole shoot). All samples were dried, ashed and extracted with 0.5 N HCI. The extract was analyzed for B and Mo by the colourimetric method and for Cu, Mn, Zn by atomic absorption spectrophotometry.

The joint analysis of hierarchical variances were conducted for the initial and residual yield effects (4 crops x 4 micronutrients rates x 4 fields x 4 replicates = 256 treatments). The joint analysis of factor variances were conducted for the initial effects of micronutrients in the first and second crop rotations (2 crop rotations x 4 micronutrient rates x 3 fields x 4 replicates = 96 treatments). The marketable yields of plants were converted to grain equivalent units (GEU). The Diagnosis and Recommendation Integrated System (DRIS) method was evaluated utilizing 10 data sets of N, P, K, Ca, Mg, B, Cu, Mn, Mo, and Zn foliar analyses to assess the state of nutrient balance of the plants (Beaufils 1973).

RESULTS

There was no evidence for any plant or year of visual symptoms of micronutrient deficiency or of changes in plant morphology and toxicity, as a result of micronutrient application. Thus, the fertilizer responses were evaluated by a comparison of yields from replicated plots and crop rotations and by comparisons of plant analysis results to critical nutrient values.

Yields of the first crop rotation

The application of boron at rates of 1, 2 and 4 kg/ha significantly increased the sugar beet yields and the yields of almost all successive crops (Table 2). The only exception was the first level of boron in maize. The average yields of the crop rotation were 57.0, 60.7, 62.6, 63.2 dt GEU/ha. Thus, yields increased significantly in the crop rotation at all B rates. The order of yield increase was the following: beet > peas > wheat > maize.

The copper fertilization at rates of 5, 10 and 20 kg/ha significantly increased the yields of the crop rotation from 54.1 in the control to 57.8, 58.3, 58.9 dt GEU/ha in the treatments, respectively. There were no significant effects of the levels applied on yields. The same reaction was obtained for all plants of the crop rotation (Table 2). The order of yield increase was beet > wheat > peas > maize.

Molybdenum application rates of 0.5, 1.0, 2.0 kg/ha resulted in significant yield increases of sugar beet and maize but only with the higher dose of molybdenum (Table 2). The crop rotation yields were increased significantly for the two higher Mo application and ranged from 56.5 in the control to 59.9, 61.6 dt GEU/ha in the treatments, respectively. There was no significant effect of the level of application on yields.

Adding <u>zinc</u> at rates of 5, 10, 20 kg/ha significantly increased the yield of maize and a higher dose of Zn resulted in the same effects on winter wheat and sugar beet (Table 2). The effects of Zn application were significant in the crop rotation for all doses (57.8, 61.3, 62.2, 63.1 dt GEU/ha) but there was no effect of the levels of fertilization. The same was true for the individual crops in the rotation.

Initial effects of micronutrients in the first and second crop rotation

The plant yields in the second rotation were higher than in the first rotation with the exception of maize (Table 3). The additional yields due to micronutrient fertilization were also higher in the second rotation. Maximum yields were achieved in both rotations where higher rates of Cu, Mo, and Zn had been applied. However, the lower rates of B, Cu and Mo were enough to bring about significant yield increases in the second rotation, with no further significant response to the higher rates. These results indicated that both secondary effects of the initial application and fresh effects of fertilization were responsible for plant yields. The only exception was the reaction of maize, where only fresh zinc supply and no secondary effects were the major determinant of yields (Table 3).

Micronutrient concentration in plant tissues

The application of micronutrients increased the micronutrient content of almost all plants and did so for almost each micronutrient addition (Table 4). During all years and for almost each plant the tissue Mo levels increased most sharply. The plant concentrations of molybdenum induced in these experiments at 1 and 2 kg Mo/ha rates were much higher than those required for plant growth. The decline in the availability of micronutrient fertilizers with the duration of soil—micronutrient contact was evident only for zinc through the decline in zinc concentration in beet (Table 4). The relative concentrations of Zn in beet tissues were 100, 84, 86, 94% at rates of 0, 5, 10 and 20 kg Zn/ha respectively.

Nutritional status of plants in the first crop rotation

The mean DRIS indices suggested that the micronutrient status was restrictive for the growth of some plants (Table 5). The negative indices in the control indicated imbalance or deficiency of the following micronutrients: B for beet and peas, Cu for wheat and peas, Zn for maize, peas and wheat.

Application of B at a rate of 2 kg/ha provided an adequate boron supply for plants in the crop rotation. If it is assumed that the adequate nutrient concentration is the concentration at which the DRIS index changes from negative to positive, then the adequate B content was 47.1 mg/g for beet and 22.2 mg/g for peas (Tables 4 and 5).

The copper rates were too low to achieve an adequate Cu supply of plants. Wheat and peas needed more than 4,5 and 6,9 (mg Cu/g in plant tissues for a non-restrictive growth (Tables 4 and 5).

Zinc concentration in maize tissues grown on a soil to which 10 kg Zn/ha had been added was 24/mg Zn/g. Such concentrations gave an adequate Zn supply for the growth of maize plants (Tables 4 and 5). However, the zinc rates were too low to obtain plant balance of Zn in crop rotation. Peas and wheat needed more than 41.0 and 22.3 mg Zn/g in plant tissue for a non-restrictive growth.

The plant yield increase found in the treatments with positive DRIS indices was probably an effect of nutrient interaction. It is not the purpose of this work to specify this interaction.

DISCUSSION

The residual effects of micronutrient fertilizers, at the investigated rates and under similar conditions should be observed 2 to 3 years for B, 4 to 5 years for Cu, 2 to 3 years for Mo and 2 to 5 years for Zn after the initial micronutrient application (Bergmann 1976). The results obtained in some field experiments may suggest that on many soils one application of Cu, Mo and Zn fertilizers is providing residual effects for many years (Murphy and Walsh 1972). However, it is difficult to find out just how long lasting this residual effect might be. For the same reason it is difficult to determine the timing and the frequency of reapplication of micronutrient fertilizers. On the other hand it is known that the contact of micronutrients with the soil reduces the effectiveness of these fertilizers over time (Barrow et al 1985, Brennan et al. 1986a, b).

The present results provide field evidence for the initial and residual values of micronutrient fertilization in crop rotation. The results obtained seem to suggest that the model of fertilization used in this study may be a simple way to correct the micronutrient status in the ordinary Polish farming practice. The application of micronutrients at the following rates: B - 2 kg/ha, Cu - 10 kg/ha and Mo - 0.5, kg/ha remains effective to correct subclinical deficiencies in crop rotations. The regular use of these rates provides secondary and fresh effects in the next rotation.

The effectiveness of applied zinc at rates of 5-20 kg/ha decreased in subsequent years. This decrease was rapid between crops 2 and 4 and Zn would again be limiting the growth in the next crop rotation. The results obtained may appear to contrast with those obtained from pot and some field studies (Murphy and Walsh 1972). Under our conditions one application of zinc fertilizer appears to provide sufficient Zn for plant growth only for two successive crops.

It has been suggested that DRIS evaluation of nutrient status of the plant is more sensitive than the critical level method. However, the order of nutrient requirement obtained by both methods was approximate.

REFERENCES

1. Barrow N.J., Leahy P.J., Southey I.N., Purser D.B. 1985.
    Initial and residual effectiveness of molybdate fertilizer in
    two areas of south western Australia. Aust.J. Agric. Res., 36,
    579-87.

2. Beaufils E.R., 1973. Diagnosis and Recommendation Integrated
    System (DRIS). A general scheme for experimentation and
    calibration based on principles developed from research in
    plant nutrition. Soil Sci. Bull. No.1, Univ. of Natal, S.Africa

3. Bergmann W. 1983 Ernahrungsstörungen bei Kulturpflanzen. VEB
    Gustav-Fischer-Verlag, Jena 1983.

4. Bergmann W. 1976. Metodiceskie ukazanija po opredeleniju
    diapazpnov vnesenija mikroelementov v pocvu. Jena.

5. Brennan R.F., Gartrell J.W., Robson A.D. 1986a. The decline in
    the availability to plants of applied copper fertilizer.
    Aust.J.Agric.Res., 37, 107-113.

6. Brennan R.F., Robson A.D., Gartrell J.W. 1986b. The effect of
    successive crops of wheat on the availability of copper
    fertilizers to plants. Aust.J.Agric.Res., 37, 115-124.

7. Gembarzewski H. 1986 Presence and deficiencies of secondary and
    trace elements in arable and grassland with high fertilizer use
    and varying soil conditions (manuscript).

8. Gupta U.C., Cutcliffe J.A. 1984. Effects of applied and
    residual boron on the nutrition of cabbage and field beans.
    Can.J.Soil Sci., 64, 571-576.

9. Gupta U.C., Cutcliffe J.A. 1982. Residual effect of boron
    applied to rutabaga on subsequent cereal crops. Soil Sci. 133,
    155-159.

10 Murphy L.S., Walsh L.M. 1972. Correction of micronutrient
    deficiencies with fertilizers. In. Mortvedt J.J., Giordano
    P.M., Lindsay W.L., eds., Micronutrient in Agriculture pp.
    347-387, Medison, Wisconsin: Soil Sci. Soc.Amer.

11. Singh M.V., Abrol I.P. 1985 Direct and residual effect of
    fertilizer zinc application on the yield and chemical
    composition of rice - wheat crops in an alkalic soil.
    Fertilizer Res., 8, 179-191.

Table 1

Characteristics of the Experimental Field

Characteristics	Data
Location	Poland, = $52^o35'$, = $16^o38'$
Soil classification	Brown soil developed from loam
Soil texture (0–20 cm)	Loamy sand
Content of humus in %	1.09
Content of clay < 2 mm in %	5.0
Soil pH (0.1 N KCI)	6.2
Micronutrient content in /mg.g^{-1}:	
Boron (HWS, Berger-Truog)	0.34 (high)
Copper (2% HNO_3, Westerhoff)	2.80 (medium)
Molybdenum (0.275M/HN_4)2^C24	
0_2O_4 – pH 3.3, Grig)	0.10 (medium)
Zinc (0.1% HCI)	22.7 (high)
Drainage	Well-drained
Mean precipitation in mm.yr^{-1}	529
Water economy	Based on rainfall
Water deficit (April–October)	from 124 to 393, mean
in mm	= – 31.4

Table 2

Initial and Residual Effects of Micronutrients

Crop	Yields (dt GEU/ha) for fertilizer levels				LSD
	0	1	2	3	P=0.05
Boron					
Sugar beet[x]	98.1	6.7	12.0	16.0	
Maize	39.5	1.8	2.6	2.9	
Peas	31.3	2.8	4.5	2.9	2.36
Winter wheat	59.1	3.3	3.4	2.9	
Copper					
Winter wheat[x]	46.7	3.5	4.0	5.5	
Sugar beet	93.6	5.1	4.8	5.8	
Maize	44.4	3.2	3.4	3.8	2.29
Peas	31.7	2.9	4.6	4.1	
Molybdenum					
Peas[x]	27.3	2.0	2.9	2.6	
Winter wheat	48.4	2.1	2.9	3.3	
Sugar beet	109.7	4.3	6.6	9.2	5.37
Maize	40.7	5.2	5.6	5.0	
Zinc					
Maize[x]	47.0	4.8	5.6	5.8	
Peas	26.8	2.7	2.9	2.9	
Winter wheat	53.5	4.0	4.4	4.2	4.22
Sugar beet	103.6	2.8	4.7	8.6	

x   = micronutrient applications
1...3 = additional yields for fertilizer levels

Table 3

Initial Effects of Micronutrients in the First and Second Crop Rotations

Crop	Crop rota-tions	Yields (dt GEU/ha) for fertilizer levels				LSD P=0.05
		0	1	2	3	
			Boron			
Sugar	1	93.1	5.4	8.4	12.6	
beet	2	100.4	9.1	5.4	8.2	4.76
			Copper			
Winter	1	45.6	3.0	3.3	4.8	
wheat	2	62.1	4.3	6.6	6.3	1.49
			Molybdenum			
Peas	1	25.1	0.9	1.9	1.3	
	2	31.1	3.7	3.7	4.5	1.00
			Zinc			
Maize	1	50.8	4.2	5.0	4.6	
	2	42.3	5.2	7.5	9.3	1.15

x = mean for three years

Table 4

Effects of a Single Application of Micronutrient
Fertilizers on Micronutrient Concentration in Plant Tissues

Crop	Mean concentration (mg/g) for fertilizer levels				Exper. adequate content[2]	Adequate content[1]
	0	1	2	3		
Boron						
Sugarbeet	39.9	38.4	47.1	51.6	47.1	35/40...10
Maize	6.3	8.6	8.7	10.7		6.....15
Peas	15.3	22.2	21.7	24.3	22.2	30.....70
Winter wheat	2.6	3.2	3.7	3.9		6.....12
Copper						
Winter wheat	3.6	4.2	4.2	4.5	> 4.5	7.....15
Sugarbeet	5.4	6.0	5.8	6.4		7.....15
Maize	4.6	4.7	5.3	5.7		6.....12
Peas	4.7	6.7	6.3	6.9	> 6.9	
Molybdenum						
Peas	2.0	6.3	21.5	19.3		0.10...0.3
Winter wheat	1.2	2.7	3.6	7.5		0.25...1.5
Sugar beet	0.4	0.8	1.3	2.7		0.15...0.5
Maize	1.2	0.8	1.2	2.2		
Zinc						
Maize	19.0	20.0	24.0	28.0	24.0	25.....70
Peas	39.0	35.0	41.0	40.0	> 41.0	
Winter wheat	21.5	22.5	22.0	22.3	> 22.3	25......70
Sugar beet	49.0	41.0	42.0	46.0		20......80

1/ from Bergmann 1983
2/ see Table 5

Table 5

Nutrient Balance of Plants by the DRIS Method

Crop	Mean DRIS indices for fertilizer levels			
	0	1	2	3
Boron				
Sugar beet	−73	−39	20	58
Maize	476	805	859	1071
Peas	−78	11	4	57
Winter wheat	182	118	128	125
Copper				
Winter wheat	−127	−32	−134	−134
Sugar beet	67	110	97	131
Maize	90	89	187	255
Peas	−130	−41	−99	−74
Molybdenum				
Peas	44	406	1910	1473
Winter wheat	73	156	140	103
Sugar beet	9	188	311	779
Maize	1	−253	−100	199
Zinc				
Maize	123	−98	21	101
Peas	−17	−173	−53	−62
Winter wheat	−33	−47	−55	−25
Sugar beet	65	4	11	32

EFFETS INITIAUX ET RESIDUELS DES OLIGO-ELEMENTS
DANS LA ROTATION DES CULTURES

M. A. Faber, Institut de recherche sur les sciences
du sol et les cultures, Pulawy

RESUME

On a procédé pendant six ans à des expériences sur le terrain
dans la région occidentale de la Pologne (latitude = $52°35$,
longitude = $16°38$) pour déterminer les effets initiaux et
résiduels des oligo-éléments appliqués dans un système de rotation
des cultures. Un oligo-élément approprié a été appliqué à une
culture particulièrement sensible à cet élément : le bore à la
betterave sucrière, le zinc au maïs, le molybdène au pois et le
cuivre au blé d'hiver. On a examiné l'effet résiduel d'une première
application d'un oligo-élément donné dans une rotation des cultures
qui s'étendait sur trois années de suite. Après la quatrième
récolte, l'oligo-élément a été à nouveau appliqué sur la même
parcelle afin de comparer les effets initiaux d'une application
précédente (première rotation) à l'application courante (deuxième
rotation). On a ensuite mesuré l'effet de l'application des
oligo-éléments sur les rendements et leur concentration dans les
plantes.

Les résultats de cette étude indiquent qu'une seule application
d'un oligo-élément dans une proportion de 2 kg/ha pour le bore, de
10 kg/ha pour le cuivre et de 0,5 kg/ha pour le molybdène corrige
efficacement la carence en ces oligo-éléments pendant toute la durée
de la rotation. Ces taux sont normaux pour les cultures de cette
région sensibles à une carence en oligo-éléments, par conséquent, il
n'y a pas de risque que l'élément appliqué rende les sols toxiques
pour les cultures suivantes. L'efficacité du zinc, appliqué dans des
proportions de 5-20 kg/ha a diminué dans les années qui suivirent.
En Pologne, il semble qu'une seule application de zinc fertilisant
fournisse suffisamment de zinc pour assurer la croissance de la
plante seulement pour deux récoltes successives.

PROBLEME DES MICROELEMENTS DANS LA
CULTURE INTENSIVE EN ROUMANIE

Mme Leticia Tiganas,
Institut de recherches sur la pédologie,
et l'agrochimie, Bucarest, Roumanie

Les observations toujours plus fréquentes et dans des zones
géographiques de plus en plus larges de certains troubles de
nutrition attribués aux oligoéléments ont stimulé le déroulement sur
un large front des recherches concernant la place et le rôle de ces
éléments dans la nutrition minérale des plantes, ainsi que l'analyse
de certains aspects d'intérêt pratique concernant les possibilités
de prévenir et de corriger les déséquilibres qu'ils provoquent.

D'excellentes synthèses relatives au stade des recherches dans
le domaine mentionné ont été publiées surtout après 1950 (25) (28)
(26) (17) (40).

A présent, on accepte unanimement le fait que dans les
conditions de la pratique d'une agriculture de type intensif qui
implique l'utilisation de quantités accrues d'amendements et
d'engrais industriels de plus en plus concentrés et purs,
simultanément à la réduction du poids des engrais naturels dans le
système de fertilisation, à l'extension de l'irrigation, à
l'application en culture de certaines variétés et hybrides, à
performances productives de plus en plus élevées et implicitement à
consommations plus grandes d'éléments nutritifs, s'accroîtra
considérablement la probabilité de dérèglements de la nutrition en
oligoéléments et par conséquent la nécessité d'intervenir avec des
moyens appropriés pour prévenir ou corriger de tels phénomènes.

En premier lieu, sont visés les sols à réserves natives
réduites en oligoéléments et/ou à conditions physico-chimiques qui
limitent la mobilité ou l'accessibilité des oligoéléments pour les
plantes; mais graduellement peuvent être affectés aussi les sols
mieux pourvus en oligoéléments où l'on a appliqué sur des périodes
plus prolongées des technologies intensives de culture. Dans de
telles situations les essais effectués dans différentes régions
géographiques du globe ont démontré que l'application d'engrais à
oligoéléments peut se solder par des gains importants de récolte et
par l'amélioration de sa qualité.

La Roumanie, comme tant d'autres pays du monde, est engagée à présent dans un ample processus de modernisation et d'intensification de la production agricole. C'est dans ce contexte que la recherche agro-chimique de Roumanie est confrontée aussi avec le problème de l'utilisation des engrais à oligoéléments dans des conditions d'efficience économique et de garantie de la protection du milieu ambiant. Toute évaluation de la dimension de ce problème doit partir, en premier lieu, de la connaissance de l'état d'approvisionnement en oligoéléments signalés dans le teritoire.

Etat d'approvisionnement en oligo-éléments des sols agricoles

On a signalé des recherches relatives aux oligo-éléments dans les sols de Roumanie, mais plus systématiquement seulement après 1950. La répartition des oligo-éléments dans les sols zonaux et intrazonaux, ainsi que leur distribution par profil ont fait l'objet de nombreuses recherches entreprises dans différentes zones, spécialement dans les parties de l'est et du sud du pays. Les résultats de ces recherches ont été récemment publiés dans des ouvrages de synthèse (6) (35).

Les sols agricoles de Roumanie ont dans leur majorité des réserves importantes d'oligo-éléments héritées du matériel parental (où prédominent les loess et les dépôts loessoides, les limons, les argiles, les dépôts alluviaux et proluviaux de différentes textures). On rencontre des réserves plus réduites d'oligoéléments dans les psammosols et dans certain sols acides à texture plus grossière (tab. 1).

L'approvisionnement en formes mobiles d'oligo-éléments n'est pas en général dans une étroite dépendance de la teneur totale en oligoéléments du sol (à l'exception des psammols), car il est déterminé par certaines propriétés physico-chimiques qui contrôlent leur mobilité chimique et par conséquent leur accessibilité pour les plantes (pH, carbonates, humus, texture, régime aéro-hydrique, etc.).

La plus grande partie des sols agricoles sont bien approvisionnés en formes mobiles d'oligo-éléments (tab. 2). Les teneurs en oligo-éléments qui dans des conditions d'agriculture intensive peuvent devenir critiques pour le développement des plantes se rencontrent sur des surperficies relativement limitées dans certaines catégories de sols.

Des teneurs réduites de formes mobiles de Zn et Mn (au-dessous
de 1,2 ppm Zn et 20 ppm Mn) ont été constatées sur de nombreux sols
châtains et carbonatiques du sud-est du pays, sur la majorité des
psammosols ainsi que sur les régosols et certain sols érodés. Les
sols halomorphes ont de même des teneurs très réduites en Mn actif
(< 15 ppm). Sur certains sols acides à texture grossière la teneur
en Zn et Mn sous des formes mobiles peut devenir critique après le
chaulage. L'utilisation de certaines doses élevées d'engrais
minéraux (en système intensif de culture) sur des sols plus
faiblement fournis en Zn et Mn peut améliorer substantiellement la
mobilité Mn (spécialement ceux à N) et peuvent déterminer chez
certaines espèces plus sensibles une nutrition déficitaire en Zn
(carence de Zn induite par P).

On rencontre des teneurs réduites de B soluble dans les
psammosols et les sols alluviaux à texture grossière (0,3 ppm),
ainsi que sur la majorité des sols lessivés (0,3 ppm), où les
conditions de nutrition en B sont satisfaisantes grâce à la réaction
acide du sol, mais peuvent devenir critiques après le chaulage.

L'approvisionnement en Mo accessible est insatisfaisante
(IMo 6,2) sur la majorité des sols acides et sur certains sols à
texture grossière. Sur ces sols, mais aussi sur des sols à faible
réaction acide jusqu'à neutre, mieux approvisionnés en Mo,
l'utilisation des doses élevées d'engrais à N peut provoquer chez
certaines cultures une carence secondaire en Mo (une insuffisance
relative par rapport aux quantités élevée de N minéral de la plante,
dont la métabolisation Mo joue un rôle essentiel par l'intermédiaire
nitrate réductase).

La richesse en Cu est en général bonne à l'exception de
certains psammosols (à réserves réduites) et certains sols
organiques (à cause de la formation de certains complexes organiques
à mobilité réduite).

Il existe des conditions de toxicité en Cu dans les vignobles
après arrachage pour reconversion, où le pH du sol est au-dessous
de 6.

Dérèglements de la nutrition en oligo-éléments et conditions de manifestation

Les dérèglements de la nutrition en oligo-éléments dans les
cultures agricoles ont été signalées et identifiés en Roumanie, les
deux dernières décennies, sur certaines surfaces de sols
caractérisés par des conditions d'accessibilité plus réduite pour

certains oligoéléments (fig. 1). Le risque et la fréquence de
manifestation de ces dérèglements s'est accrue avec la promotion
toujours plus ferme de certaines méthodes intensives de culture du
sol.

La carence en zinc est la plus répandue des déficiences de la
nutrition en oligo-éléments signalées. Les premières observations
sur la carence en Zn chez le maïs ont été faites dans la Plaine du
Baragan et dans le Nord de la Moldavie. (32) (29) (18).

A présent la carence en zinc affecte spécialement la culture du
maïs et en moindre mesure celles du haricot, du soja et de certains
arbres fruitiers (pommier). Le phénomène se manifeste avec
prépondérance dans la zone des chernozems, des sols châtains et sur
les sol érodés ainsi que sur certains psammosols. Parmi les
facteurs de risque de ce dérèglement identifié en Roumanie on peut
mentionner les suivants : teneur réduite en zinc accessible du sol
(inférieure à 1 ppm); réaction neutre et faiblement alcaline des
sols et présence de carbonates dans l'horizon supérieur; texture
grossière du sol (réserve réduite de Zn); fertilisation régulière à
doses élevées d'engrais à N et spécialement à P; décapages lors du
nivellement capital du sol pour irrigations; extension des hybrides
tardifs; monoculture du maïs; stagnation de l'eau sur le sol au
semis; temps froid et humide au départ de la végétation, qui
accentue en grande mesure l'action négative des autres facteurs de
risque mentionnés antérieurement.

Stimulés par la fréquence et l'ampleur du phénomène de carence
en zinc ainsi que par leurs conséquences économiques, on a déployé
ces dernières années en Roumanie, sur un large front, des recherches
agrochimiques qui ont élargi la zone de connaissance dans ce domaine
et ont contribué en même temps à la solution de certains aspects
d'intérêt pratique.

Pour pronostiquer le phénomène de carence en zinc chez le maïs
et le haricot on a proposé plusieurs schémas qui ont en vue les
principales propriétés agrochimiques impliquées dans ce
déséquilibre, (Zn accessible, pH du sol, P accessible) (9). Sur le
tableau 3 sont présentées deux possibilités de pronostic sur la base
de "l'indice de carence en zinc" (ICZn) et sur la base de "l'indice
réaction-phosphates mobiles" (IRPM) qui peuvent être utilisées en
fonction des données agrochimiques disponibles. Leur élabortion a
été faite sur la base de la corrélation établie par les observations
de terrain, entre la manifestation du phénomène et les propriétés
respectives du sol. La facon de réunir les propriétés agrochimiques
se base sur la connaissance de l'influence de ces propriétés sur
l'absorption du zinc par les plantes.

Les recherches effectuées sur la carence en Zn ont montré que le mécanisme d'apparition de ce trouble est complexe et pourrait être causés par les rythmes différents où s'absorbent dans la plante les éléments nutritifs pendant la période critique de manifestation du phénomène. Les données expérimentales ont montré que l'accumulation Zn dans les plantes a lieu à un rythme plus lent que celle du P, N et K pendant cette période, ce qui détermine des valeurs plus réduites des rapports entre Zn et macroéléments. Les niveaux critiques de ces rapports (au-dessous desquels les symptômes deviennent manifestes) caractéristiques de la carence en Zn, diffèrent d'une espèce à l'autre et sont en étroite relation avec les propriétés du sol, conditions climatiques et facteurs agrochimiques qui accroissent l'absorption du P et réduisent l'absorption du Zn dans la plante (8). Lorsque ces rapports atteignent des seuils critiques, il se produit dans les plantes des déséquilibres profonds, se manifestant par la baisse de la concentration en Zn et l'augmentation de la concentration en P, Mn, Fe et en moindre mesure en Cu,, et entre ces éléments s'établissent des interactions antagoniques (Zn-P, Zn-Mn; Zn-Fe, Zn-Cu), ou synergiques (P-Mn; P-Fe; P-Cu) (tabl. 4). Ces interactions spécifiques auxaffectées par la carence en Zn ne se manifestent pas dans des situations normales de nutrition. Le déséquilibre de la balance entre éléments de nutrition dans le cas de carence en Zn a été signalé dans des ouvrages publiés en Roumanie et dans d'autres pays (1) (5) (36) (39).

La prédisposition des variétés hybrides à la carence en Zn, contrôlée probablement génétiquement, semble être toujours une conséquence des rythmes différents où s'absorbent et s'accumulent dans les plantes certains éléments nutritifs (Zn, P,, Mn, Fe) pendant la première partie de la végétation (34). On a élaboré, sur la base de ces recherches, un procédé pour établir la susceptibilité relative des hybrides de maïs à la carence en Zn, où on utilise une série d'indices formulés sur la base des rapports entre les éléments de nutrition dans la partie aérienne des plantes pendant la période 3 et 8 à 10 feuilles. L'arrangement des hybrides (cultivés sur le même sol et agrofonds de fertilisation) selon leur prédisposition à la carence en Zn se fait d'après la somme des indices ou selon la position moyenne établie par rapport à chaque indice utilisé (tabl. 5) (10).

La carence secondaire en molybdène a été signalée ces dernières années chez le tournesol, le maïs et d'autres cultures de plein champ, sur une large gamme de sols à réaction acide jusqu'à neutre, dans la situation d'application de doses élevées d'engrais azotés lors de la préparation du lit germinatif (2) (22). On a mis en évidence dans les plantes affectées, des corrélations étroites entre la teneur en nitrates et l'activité nitrate réductase. Sur certains sols le phénomène a été associé à la déficience en P, K et S et à la toxicité de Al et Mn (24).

La carence en B ont été mis en évidence par des symptômes spécifiques chez le pommier (fruits) dans des plantations emplacées sur psammosols de l'Ouest du pays en régime non-irrigué, ainsi que sur certains sols acides fortement débasifiés du nord-ouest du pays, faiblement approvisionnés en bore soluble (< 0,2 ppm) (4). Le chaulage excessif des sols acides ou l'application non-uniforme des amendements, sont des facteurs de risque importants sur ces sols. La déficience en B a été observée chez la même espèce aussi sur les psammosols nivelés et modelés des terrases du Jiu, mais à intensité moindre due à l'irrigation (16). En général, le phénomène a été plus intense pendant les années sèches et dans les conditions d'application de doses élevées d'engrais potassiques et azotés.

Les années à précipitations plus réduites pendant la seconde moitié de juillet et la première moitié d'août, dans les cultures non-irriguées de betterave sucrière sur chernozems et sols alluviaux à réaction neutre faiblement alcaline, on a signalé relativement fréquemment des symptômes spécifiques à la déficience en B ("maladie du coeur"). On a rencontré des symptômes de carence en B chez le tournesol les années à printemps secs, sur des sols acides chaulés et sur des sols lourds faiblement fournis en bore hydrosoluble (< 0,3 ppm).

La carence en fer (chlorose ferrique) se manifeste fréquemment, dans les plantations de pêcher et vigne sur régosols, sols châtains, chernosems carbonatiques.

Des recherches systématiques effectuées chez le pêcher (15) et la vigne (30) ont mis en évidence un complexe de facteurs à rôle déterminant dans la manifestation du phénomène : (réaction alcaline et présence du calcaire actif, régime aérohydrique défavorable, compactage du sol ou drainage défectueux, application systématique de doses élevées d'engrais organiques et minéraux, spécialement au phosphore, porte-greffes génétiquement inefficaces dans l'absorption du fer sous formes plus difficilement accessibles.

On n'a pas identifié jusqu'à présent de phénomènes de carence en Mn, mais sur certains sols salins et alcalins on a obtenu expérimentalement des effets favorables à la fertilisation au Mn (23).

Des phénomènes de toxicité au Mn ont été signalés dans les plantations fruitières (pommier-variétés Starking et Starkrimson) situées sur sols acides, surtout celles affectées par des processus de pseudogleysation à teneurs en Mn actif dans le sol dépassant 120 ppm, dans la situation de la fertilisation systématique à azotate d'ammonium (16). De même, on a signalé des intoxications au Mn, sur psammosols cultivés intensément, chez le pêcher et certaines cultures annuelles (maïs, soja, luzerne). Les concentrations de Mn réalisées dans les plantes affectées ont varié en fonction de l'espèce (> 300 ppm chez le pêcher, > 800 ppm chez le maïs (3) (16).

Le phénomène est prévenu par l'application contrôlée des engrais azotés et l'utilisation de sources d'azote non-acidifiantes, le chaulage dans le cas des sols acides et le drainage de l'excès d'eau dans le cas de ceux pseudogleysés.

La carence en cuivre ne s'est pas manifestée jusqu'à présent par des symptömes spécifiques, mais les états végétatifs plus mauvais rencontrés souvent chez les céréales à chaume sur certains sols tourbeaux (formés spécialement dans des conditions d'oligotrophisme) récemment améliorés et mis en culture peuvent être en relation avec ce phénomène.

On a encore constaté des teneurs réduites en cuivre chez les céréales et des rapports déséquilibrés N/Cu sur des psammosols, irrigués, cultivés intensément ainsi que sur des sols normalement fournis en cuivre chez les variétés très productives, pendant certaines années climatiques. (3) (37).

Possibilités et moyens de prévenir et de traiter les déficiences en oligoéléments

L'expérience accumulée sur plan mondial sur l'utilisation des engrais à oligoéléments est vaste et précieuse. Jusqu'à présent se sont dessinées deux directions principales d'intervention avec ces types d'engrais.

interventions préventives : application d'engrais (simples ou complexes) dans le sol (par épandage ou en bande) efficaces sur des durées variables de temps; application de préparations à oligoéléments sur la semence (par arrosage, poudrage ou trempage);

- interventions curatives : traitement foliaire des cultures affectées avec des solutions diluées de sels des oligoéléments ou à engrais foliaires complexes à macro et oligoéléments.

Les doses d'engrais à oligoéléments pour l'application préventive ou curative ont été établies en général sur des bases expérimentales. Elles varient en fonction de la nature de l'oligoélément impliqué, de l'état d'approvisionnement du sol pour l'élément concerné, des exigences des plantes cultivées, du degré d'intensification de la production végetale, des conditions climatiques etc. Utilisées irrationnellement, elles peuvent provoquer des phénomènes de phytotoxicité.

En Roumanie, les doses d'engrais à oligoéléments, les méthodes techniques d'application, l'efficacité et la durée de leur effet ont été établies dans des champs d'essais, spécialement pour le Zn, B et Mo. Par exemple, l'application préventive dans le sol de 7 à 10 kg Zn/ha sur des sols à ICZn sous 3,4 a donné chez le maïs des gains de récolte de 10 % - 20 %. Dans le cas des traitements foliaires sur des plantes à carence, le gain de récolte pour 1 kg Zn s'élève à 1500 - 2000 kg de maïs grains (13).

Les recherches expérimentales ont servi à l'élaboration de certains modèles mathématiques pour établir les doses d'engrais oligoéléments, qui englobent tant les facteurs déterminants des carences primaires que ceux qui provoquent ou favorisent les carences secondaires. Sur les tableaux 6, 7 8 sont présentés ces modèles et les doses optimales scientifiques (DOS) d'oligoéléments (Zn, B et Mo) calculées pour les plantes de plein champ. Les doses recommandées sont destinées aux interventions préventives et poursuivent la création dans le sol de teneurs normales et durables d'oligoéléments accessibles (11). Pour les plantations fruitières et viticoles les doses sont corrigées en fonction des données de l'analyse foliaire.

Les doses sont plafonnées aux niveaux maximum de 9 kg Zn/ha, 2kg B/ha et respectivement 1 kg Mo/ha pour les sols faiblement fournis en ces éléments et baissent progressivement jusqu'à annulation chez les sols moyens et bien fournis. Leur application dans le sol assure le nécessaire en oligoéléments des cultures pour au moins 4 à 5 ans.

Chez les cultures très intensives, même sur des sols bien fournis en oligoéléments, s'est avérée très efficace la fertilisation supplémentaire avec des préparations spéciales de stimulation appliquées sur les feuilles, qui comprennent aussi de petites quantités d'oligoéléments.

La production des semences et de matériels de plantations à teneurs plus élevées en oligoéléments, et leur utilisation, spécialement sur des sols fournis plus faiblement en oligoéléments, apparaissent de même comme une possibilité réelle de réduire l'incidence des troubles de nutrition et de promouvoir des rythmes initiaux élevés de croissance, condition essentielle pour les cultures intensives.

Sur le tableau 9 sont présentées quelques données dans ce sens, obtenues chez le maïs sur des lots d'hybridation et des lignées parentales, par fertilisation foliaire à Zn et Mo. La semence obtenue de ces traitements, semée sur des sols plus faiblement fournis en zinc, a déterminé des gains importants de récolte (5 à 34 %) sur un agrofonds optimum de N, P, K (Borlan, Tiganas, et all. données non-publiées).

Les engrais naturels, appliqués périodiquement, restent une source appréciable d'éléments nutritifs, y compris les oligoéléments, en indiquant que ceux produits dans les grands complexes d'élevage intensif sont dans la majorité des cas, grâce au mode spécifique d'affourragement, à teneur plus élevée en oligoéléments et métaux lourds (tabl. 10) ce qui impose certaines restrictions dans leur utilisation, afin d'éviter des phénomènes de pollution irréversible du sol (38).

La création par des travaux de génétique végétale de certaines variétés et hybrides plus efficients dans l'absorption et l'utilisation des oligoéléments est une voie relativement récente pour prévenir les dérèglements de nutrition en oligoéléments (20). Pour le moment, avec des possibilités réelles d'application pratique, il reste à établir par voie expérimentale ou à l'aide de certains indicateurs indirects, des types plus sensibles au dérèglement de la nutrition en oligoéléments, en vue d'éviter leur culture sur des sols à conditions potentielles de déficience. C'est dans ce sens que nous avons mentionné antérieurement le moyen d'établir et d'utiliser en Roumanie des indices de susceptibilité relative des hybrides de maïs à la carence en Zn.

Orientations et recommandations pour l'utilisation des engrais à oligoéléments

Les recherches agrochimiques réalisées en Roumanie jusqu'à présent ont réussi à dessiner assez clairement les dimensions du problème des oligoéléments dans la production végétale. On a

accumulé de nombreuses données concernant l'état d'approvisionnement
des sols en oligoéléments, sur la base desquelles on a pu identifier
les sols où sont possibles des troubles de nutrition. On a signalé
et précisé les conditions de manifestation de certaines carences en
oligo-éléments (les plus importantes comme aire de manifestation et
fréquence étant celles de zinc, molybdène et bore) : on a établi par
des essais en champ les méthodes technologiques d'application et
l'efficacité des engrais avec ces oligoéléments.

Si les quantités d'oligoéléments extraites avec les récoltes ne
sont pas restituées au sol par fertilisation, on doit s'attendre au
fur et à mesure de l'intensification de l'agriculture et de
l'augmentation continuelle des récoltes, à une sollicitation plus
intense des réserves en oligoéléments du sol et à un rythme plus
rapide d'épuisement (tab. 11).

Une série de mesures d'optimisation des propriétés physiques et
chimiques de certains sols, nécessaires à l'amélioration de leur
fertilité naturelle en vue de la culture intensive, peuvent
déterminer dans certains cas une aggravation des conditions de
nutrition avec certains oligoéléments (par ex. la baisse de la
mobilité du bore sur les sols acides chaulés, la carence en zinc
induite par la fertilisation phosphatique intense, etc.).

D'autre part, les variétés et les hybrides créés pour des
productions élevées et très élevées sollicitent du sol de grandes
quantités d'éléments nutritifs à formes facilement assimilables,
étant moins efficients dans leur extraction de formes
potentiellement accessibles (19). Cette particularité confère un
risque supplémentaire de dérèglement de la nutrition en
oligoéléments dans l'agriculture intensive.

Il est possible, dans ce contexte, que l'apparition de certains
troubles de nutrition en oligoéléments augmente en superficie et en
intensité les années suivantes, avec des conséquences négatives sur
la quantité des récoltes et leur qualité, imposant des mesures
d'intervention pour les prévenir et les corriger.

Les recommandations de fertilisation en oligoéléments, en
Roumanie, se base sur la conception que la prévention des carences
de nutrition en oligoéléments doit être réalisée, en premier lieu
par un ensemble de mesures agrophytotechniques telles que :appliquer
périodiquement des engrais naturels, éviter la culture des variétés

et hybrides les plus susceptibles sur sols faiblement fournis en oligoéléments, pratiquer des rotations adéquates des cultures, éviter la surfertilisation minérale, utiliser des semences et des matériels de plantation à teneur optimale etc. Les interventions avec engrais spéciaux à oligoéléments doivent être effectuées seulement dans des situations plus aigües où les mesures agrophytotechniques mentionnées n'ont pas eu l'efficience escomptée (12).

L'objectif du relèvement de la teneur en formes mobiles d'oligoéléments du sol à des niveaux optimum par l'application d'engrais à oligoéléments sera réalisé graduellement à l'avenir, en étroite corrélation avec les exigences de l'agriculture et avec les possibilités matérielles de l'économie nationale. Dans ce sens, dans une première étape sont visés les terrains où les années antérieures ont été fréquemment signalées des carences en oligoéléments, les cultures porte-graine et les pépinières arboricoles et viticoles (pour obtenir un matériel de reproduction vigoureux capable de résister dans de conditions de stress) et les terrains occupés par des cultures très intensives; dans une étape suivante, on a en vue tous les sols qui ne sont pas fournis au niveau optimum en formes mobiles d'oligoéléments.

Sur la base des données dont dispose à présent la recherche agrochimique dans ce domaine en Roumanie, on estime pour la période 1986-1990 les besoins annuels d'environ 600 t zinc, 1000 t bore et 15 t molybdène (13).

BIBLIOGRAPHIE

1. Adriano D.C., Paulsen G.M., Murphy L.S. 1971.. Agron J.,63 : 36-39
2. Apostol V., Lacatusu R., 1978. Prod. Veget. - Cereals si Plante Technice, 6 : 39-47
3. Bajescu I., Manuca O., Daniliuc D., Nicolescu M., 1977. An. Inst.Cercet. Ped. Agrochim. XLII : 137
4. Bajescu I., Chirriac A. 1980. An. Inst. Cercet. Ped. Agrochim.. XLIV : 65
5. Bajescu I., Chiriac A., Tiganas L. 1982. Anal. Inst. CErcet. Ped. Agrochim. XLV : 67
6. Bajescu I., Chiriac A. 1984 "Distributia microelementelor in solurile din Romania. Implicatii in agricultura" Ed. Ceres Bucuresti
7. Berger K.C., Truog E. 1940. J.Am. Soc. Agron. 32 : 397-301
8. Borlan Z., Hera Cr.,, Dornescu D., Lacatusu R., Rands D., Armeanu M. 1976, bull. de l'Acad. Sci. Agr.. Forest. 5 : 73-87
9. Borlan Z., Hera Cr., Dornescu D., Lacatusu R. 1977, Coll. Franco-Roumain. La fertilité du sol et la nutrition oligominérale des plantes, Bordeaux, p. 91
10. Borlan Z., Tiganas L., Dornescu D., Mihaila V., Lacatusu R. 1977 - Procedee de stabilire a gradului de susceptibilitate a hibrizilor de prumb la carenta de zinc. Descrierea inventiei 74090 O.S.I.M.. R.S.R.
11 Borlan Z., Hera Cr. si altii, 1982. Tabele si nomograme agrochimiice Ed. Ceres Bucuresti p. 99
12. Borlan Z., Hera Cr..1984 "Optimizarea agrochimica a sistemului sol-planta" Ed. Academiei RSR Bucuresti.
13. Borlan Z., Vintial I., Stefan O., Bunescu O., Tiganas L., Bogaci R., Ghidia A. 1986. Bul. Inform Acad. St. Agric. si Silv.. 15 : 15-27
14. Brar, S.P.S., Randhawa, N.S. Dwivedi R., 1974 in Plant Analysis and Fertilizer Problems. Proc. of 7th Intern. Colloque (Hanover) p. 55-71
15. Chiriac A. Tiganas L. 1981. dirijarea fertilizarii in plantatii de piersic. ICPA - Rapport intern.
16. Chiriac A., Tiganas L. 1982. Stabilirea criteriilor de aplicare diferentiata a ingrasamintelor in plantatii de mar. ICPA - Rapport intern.

17.   Davies E. Brian (Ed.) 1980.   Applied Soill Trade Elements.
       Ed. John Wiley and Sons.
18.   Dornescu D.,,, 1969. Simpoz. Nat. Agrochim. 7, Bucuresti, 85-95
19.   Finck A. 1982. Fertilizers and Fertilization, Ed. Verlag
       Chemie pag. 1-30.
20.   Gerloff G.G...., Gabelman, W.H., 1983. Genetic Basis of
       Inorganic Plant Nutrition, in Inorganic Plant Nutrition, Ed.
       A. Läuchli, R.L. Bieleski, Springer-Vrlag p. 472-476
21.   Gupta Umesh C., Lipsett J.,, 1981 Adv. in Agron. 34 : 80
22.   Hera Gr., Borlan Z., 1980. Ghid pentru alcatuirea planurilor
       de fertilizare, Ed. Ceres Bucuresti, 54-62
23.   Jamel A.M., 1981. Teza de doctorat, Inst. Agron. "N. Balcescu"
       Bucuresti
24.   Lacatusu R., Borlan Z., Nastase A., Manea P., Borza I. 1983
       Lucr. Conf. Nat. St. Solului, Braila, 21, B, 136-145
25.   Mitchelle R.L., 1964 "Trace elements in soils" In chemistry of
       the Soil, 2-nd edn, Ed.F.E. Bear, Reinhold N.Y.
26.   Mortvedt I.I., Giordano P.M., Lindsay W.L., (Ed.) 1972
       Micronutrients in Agriculture, Soil Sci. Soc. Amer. Inc.
       Madison, Wisconsin
27.   Müller K.H., Wuth E., Witter B.,, Ebeling R., Bergman W. 1964
       Albrecht - Thaer - Archiv., 8,p.   353
28.   Peive Ia.V. 1961. Microelementi i ih znacenie v seliskom
       hozeaistve. M., 1961.
29.   Petrovici P., Plesa D. 1968. Probleme agricole, 3, p. 15
30.   Rauta C. 1982. Teza de doctorat, Inst. Agron. "N. Balcescu"
       Bucuresti
31.   Safaya, M.M., Bhart - Singh 1974.   Plant and Soil 48 (2)
       279-290
32.   Stefan Gh., Marinov C. 1966 - Problème Agricole, 7
33.   Trierweiler J.E., Lindsay W.L. 1969.   Soil Sci. Soc. Amer.
       Proc. 33 (1) p. 49
34.   Tiganas L., Bajescu I., Dornescu D.. 1977. Coll.
       Franco-Roumain, Bordeaux, p. 165-169
35.   Tiganas L., Borlan Z., 1984. "Starea de asigurare cu micro
       elemente a solurilor" in Situatia agrochimica a solurilor
       României.. Prezent si viitor, Ed. Ceres, Bucuresti
36.   Vintila I., Bajescu I., Muller S.. 1977. Anal. Inst. Cercet.
       Ped. Agrochim. XLII p. 149
37.   Vintila I., Stefan O., Nitescu S. 1981. Rapport intern. Inst.
       Cerc. Ped. Agrochim. Bucuresti
38.   Vintial I., tiganas L., 1983. Determinarea microelementelor in
       Metode de evaluare a ingrasamintelor organice ca surse de
       substante nutritive. MAIA, Metode, rapoarte, indrumari nr. 15.
39.   Warnock R.E., 1970. Proc. Soil Soc. Amer. 34 : 765-769
40.   Zirin N.G., Sadovnicov L.K., 1985. Himia tiajelih metalov
       misiaka i molibdena v pocivah Uzd. Moskov. Univ.

Tableau 1

Teneur totale en oligoéléments (en ppm) de l'horizon supérieur
dans les principaux sols agricoles de Roumanie
(d'après Tiganas et Borlan 1984)

Types de sol	Zinc		Bore		Molybdène		Cuivre		Manganèse	
	I.V. X	M. XX	I.V.	M.	I.V.	M.	I.V.	M.	I.V.	M.
Sols châtains	55-70	62	50-55	50	-	-	18-21	19	545-715	645
Chernozems	43-102	74	42-54	48	1,0-4,6	2,5	11-38	24	450-1133	776
Chernozems cambiques et argilliques	58-97	73	28-56	42	0,4-1,0	0,8	15-36	23	420-1373	883
Sols gris forestiers	42-70	54	29-38	34	-	-	16-26	21	780-1000	841
Sols bruns rougeâtres et brun argilliques	57-83	70	31-50	39	-	-	22-39	24	675-1171	998
Sols bruns lessivés	40-79	60	28-68	43	0,4-1,6	1,0	12-33	21	523-2629	1244
Sols lessivés (luvisols albiques)	35-70	53	26-68	46	0,2-1,1	0,8	7-26	16	218-3215	1061
Psammosols (sols sableux)	11-24	20	12-18	14	0,43	-	3-5	4	175-496	309
Sols alluviaux et alluvions	39-80	60	14-21	19	0,6-2,2	1,4	20-50	31	610-900	691
Sols halomorphes	50-88	73	44-148	78	0,6-0,9	0,8	3-33	19	580-810	664

X) I.V. - intervalle de variation
XX) M. - valeur moyenne

Tableau 2

Teneur en formes mobiles [x)] d'oligoéléments (en ppm) de l'horizon supérieur dans les principaux sols agricoles de Roumanie (intervalles de variation I.V. et valeurs moyennes - M) (d'après Tiganas et Borlan 1984)

Types de sol	Zinc		Bore		Molybdène		Cuivre		Manganèse	
	I.V.	M.	I.V.	M.	I.V.	M.	I.V.	M.	I.V.	M.
Sols châtains	0,5-1,2	0,7	0,9-1,6	1,2	0,06	-	3,3-4,0	3,7	22-40	29
Chernozems	0,8-2,5	1,2	0,6-2,9	1,0	0,05-0,38	0,09	4,5-14	8,8	7-141	51
Chernozems cambiques et argilliques	0,7-4,0	1,7	0,2-1,7	0,7	0,03-0,33	0,17	2,5-21	6,7	38-218	123
Sols gris forestiers	1,1-2,3	1,5	0,2-0,7	0,5	0,13-0,33	0,23	2,3-18	6,5	78-202	151
Sols bruns rougeâtres et bruns argilliques	1,2-2,6	1,8	0,4-1,0	0,6	0,18-0,29	0,22	3,5-6,0	4,8	53-196	117
Sols bruns lessivés	0,6-2,1	1,5	0,1-0,5	0,3	0,17-0,33	0,24	2,7-7,0	4,1	42-262	108
Sols lessivés	0,8-5,2	2,1	0,1-0,4	0,3	0,02-0,73	0,26	1,3-4,0	2,4	21-269	111
Psammosols	0,2-1,1	0,5	0,1-0,4	0,2	-	-	0,4-1,1	0,7	2-31	14
Sols alluviaux	0,7-3,1	1,7	0,3-0,9	0,6	0,06-0,49	0,29	3,0-14	7,5	7-69	20
Régosols, érodisols	0,6-1,4	0,9	0,3-0,9	0,6	0,06-0,16	0,11	1,1-6,0	2,9	3-100	40
Sols halomorphes	0,8-2,8	1,5	2,0-21	7,2	0,17-0,70	0,43	3,6-8,2	6,7	2-13	7

X) Zn - EDTA + $(NH_4)_2CO_3$ (PH 8,6); Mn actif - méthode schachtshabel (PH 8); Mo - méthode Grigg

B_{H_2O} - méthode Berger Truof; Cu - Na_2EDTA (pH 7)

Tableau 3

Indices d'appréciation de la probabilité
de l'apparition de la carence en zinc
(d'après Borlan et coll. 1977)

Indice réaction-phosphates mobile IRPM	Indice de carence en Zn ICZn	Probabilité de l'apparition de la carence en zinc
0,288	1,7	grande
0,289 - 0,384	1,8 - 3,4	appréciable
0,385 - 0,576	3,5 - 5,1	moyenne
0,577 - 1,165	5,2 - 6,8	petite
1,165	6,8	très petite

$$IRPM = \frac{90 - 10 \ pH}{P_{AL}}$$

$$ICZn = \frac{Zn. \ (1,3 \ pH - 0,11 \ pH^2 - 2,82). \ 100}{P_{AL}}$$

pH - déterminé dans la suspension aqueuse 1 : 2,5

P_{AL} (ppm) - extrait dans l'acétate - lactate d'ammonium selon Egner et al. (1960)

An (ppm) - extrait dans l'EDTA + $(NH_4)_2CO_3$ selon Trierweiler et Lindsay (1969)

Tableau 4

Modifications dans la composition chimique des plants de maïs (HS 400) dans la phase 4 à 5 feuilles sous l'influence de la fertilisation à P et Zn, sur un chernozem typique (d'après Stefan et Tiganas 1981)

Traitement hg/ha		ICZn	Probalité de carence en Zn après ICZn	Aspect des plantes	Zn	Cu	Mn	Fe	P		P/Zn
P	Zn				En ppm de M.S.				%		
0	0	8,7	Très petite	Normal	17	7,9	80	177	0,18		105
120	0	4,3	Moyenne	Carence	15	12,7	159	481	0,37		250
240	0	3,0	Appréciable	Carence intense	14	12,7	168	469	0,83		590
0	20	33,4	Très petite	Normal	43	8,6	77	205	0,17		39
120	20	16,8	Très petite	Normal	37	9,3	92	216	0,29		79
240	20	11,7	Très petite	Carence faible	32	9,3	92	221	0,46		145

Coefficients de corrélation (r) (dans les phases 4 à 8 feuilles)

	P	Zn
Zn	− 0,48[0]	− 0,68[000]
Cu	+ 0,79[xxx]	− 0,75[000]
Mn	+ 0,79[xxx]	− 0,70[000]
Fe	+ 0,48[x]	

n = 24

0,x – significatif pour P 5 %

000; xxx – significatif pour P 0,1 %

Tableau 5

Susceptibilité relative de certains hybrides de maïs, appréciée à l'aide de certains indices proposés par Borlan et coll., 1977

Hybride	Indices de susceptibilité à la carence en Zn [x]			Position relative moyenne dans la série des susceptibilités
	P x Fe/Zn	P x Mn/Zn	PxFexMn/Zn	
En conditions de probabilité réduite de l'apparition de la carence en Zn (ICZn = 6,0)				
HS 218	2,9 (1)[xx]	0,3 (1)	285 (2)	1,3
HS 370	3,0 (2)	0,5 (2)	282 (1)	1,7
HS 420	4,5 (3)	0,5 (2)	381 (3)	2,7
En conditions de probabilité très grande de l'apparition de la carence en Zn (ICZn = 1,6)				
HS 218	8,7 (1)	1,1 (1)	653 (1)	1,0
HS 370	9,1 (2)	1,1 (1)	769 (2)	1,7
HS 420	13,0 (3)	1,2 (2)	1182 (3)	2,7

x) Les symboles chimiques représentent la teneur en pourcentage dans le cas P et les teneurs en ppm dans le cas Zn, Mn et Fe, dans les plantes entières de maïs dans la phase 4 à 6 feuilles.

xx) ( ) position dans la série de susceptibilité relative (1 = petite; 3 = grande).

Tableau 6

Doses optimales scientifiques (DOS) de zinc (en Kg Zn/ha) pour l'application préventive une fois à 5 à 6 ans, dans les assolements de maïs, haricot et soja en fonction de la réaction et de la teneur des formes mobiles de zinc et de phosphore dans la couche arable des sols (d'après Borlan et al. 1982)

Zn.Fr x)	P_{AL} (ppm P)									
	10	20	30	40	50	60	70	80	90	100
0,25	4,1	7,6	8,2	8,4	8,6	8,7	8,7	8,8	8,8	8,8
0,50	-	4,1	6,7	7,6	8,0	8,2	8,3	8,4	8,5	8,6
0,75	-	-	4,1	6,2	7,1	7,6	7,9	8,0	8,2	8,3
1,00	-	-	-	4,1	5,9	6,8	7,2	7,6	7,8	8,0
1,25	-	-	-	-	4,1	5,6	6,5	7,0	7,3	7,6
1,50	-	-	-	-	1,7	4,1	5,5	6,2	6,7	7,1
1,75	-	-	-	-	-	2,1	4,1	5,3	6,1	6,5
2,00	-	-	-	-	-	-	2,4	4,1	5,2	5,9
2,25	-	-	-	-	-	-	-	2,7	4,1	5,1
2,50	-	-	-	-	-	-	-	0,9	2,8	4,1
2,75	-	-	-	-	-	-	-	-	1,3	3,0

x) FR = facteur réaction = $1,3 \ pH - 0,11 \ pH^2 - 2,819$

Formule de calcul : DOS de Zn, Kg/ha = $10 - 10^{0,307 \cdot ICZn}$ ou $ICZn = \dfrac{Zn.FR}{P_{AL}}$

Tableau 7

Doses optimales scientifiques (DOS) de bore, pour l'application préventive une fois à 5 à 6 ans dans les assolements de betterave sucrière, betterave et chou-rave fourrager, lin pour huile, en fonction de la teneur en bore soluble (B H_2O). (Berger - Truog) et en argile dans la couche arable (d'après Borlan et al. 1982)

Teneur en argile (sous 0,002 mm) (Ag) %	DOS de B (Kg/ha), lorsque la teneur en B_{H_2O} (en ppm) du sol est de :							
	0,10 et sous	0,15	0,20	0,25	0,30	0,35	0,40	0,45 et au-dessus
10	0,95	0,86	0,76	0,64	0,51	0,36	0,19	0,0
15	1,41	1,28	1,14	0,96	0,75	0,53	0,28	0,0
20	1,63	1,48	1,32	1,11	0,87	0,62	0,32	0,0
25	1,77	1,61	1,43	1,21	0,95	0,67	0,35	0,0
30	1,87	1,70	1,52	1,28	1,00	0,71	0,37	0,0
35	1,94	1,76	1,57	1,32	1,04	0,73	0,38	0,0
40	1,98	1,79	1,60	1,35	1,06	0,75	0,39	0,0
45	2,03	1,84	1,64	1,38	1,09	0,77	0,40	0,0
50	2,05	1,86	1,65	1,39	1,09	0,77	0,40	0,0
55	2,08	1,89	1,68	1,42	1,11	0,79	0,41	0,0

Formule de calcul: DOS de B, Kg/ha = $(3-3 \ 2,22 \ B_{H_2O}) (1,35 - \frac{8}{Ag})$

$(1,35 - \frac{30}{DOE \ N})$

où : DOE N = dose (optimale économique) d'azote nécessaire à la culture de l'assolement où on a appliqué le bore

Tableau 8

Doses optimales scientifiques (DOS) de molybdène pour l'application préventive une fois à 5 à 6 ans dans les assolements à légumineuses pérennes (trèfle, luzerne, lotier corniculé) et annuelles (soja, pois, pois chiche, haricot), tournesol, pomme de terre, betterave sucrière, en fonction de "l'indice de molybdène" IMo et de la teneur en argile dans la couche arable (d'après Eorlan et al. 1982)

Teneur en argile (sous à 0,002 mm) (Ag) %	Dose de Mo (Kg/ha) lorsque IMo est de :								
	5,50	5,75	6,00	6,25	6,50	6,75	7,00	7,25	7,50
10	0,46	0,41	0,34	0,26	0,18	0,12	0,08	0,05	0,03
15	0,68	0,61	0,51	0,39	0,27	0,18	0,11	0,07	0,04
20	0,79	0,70	0,58	0,45	0,31	0,21	0,13	0,08	0,05
25	0,86	0,76	0,63	0,49	0,34	0,27	0,14	0,08	0,05
30	0,91	0,81	0,67	0,51	0,36	0,24	0,15	0,09	0,05
35	0,94	0,84	0,69	0,53	0,38	0,25	0,15	0,09	0,06
40	0,96	0,85	0,72	0,54	0,38	0,25	0,16	0,09	0,06
45	0,98	0,87	0,73	0,56	0,39	0,26	0,16	0,10	0,06
50	0,99	0,88	0,74	0,56	0,40	0,26	0,16	0,10	0,06
55	1,01	0,89	0,75	0,57	0,40	0,27	0,17	0,10	0,06

Formule de calcul : DOS Mo, Kg/ha = $(1,35 - \frac{8}{Ag})$ $(1,25 - \frac{18}{DOEN})$ $(\frac{10 - IMo}{10 - 6,2 + 10 - IMo})$

Où DOEN = dose (optimale économique) d'azote nécessaire à la culture de l'assolement où on applique les engrais à molybdène.

Tableau 9

Influence de certains traitements foliaire x) appliqués dans les
lots d'hybridation et lignées parentales de certains hybrides de
maïs sur la teneur en oligoéléments (Zn, Mo) de la semence et sur
la production de grains obtenus de la semence enrichie en Zn et Mo.
Résultats obtenus dans les champs d'essais situés sur des chernozems
et des chernozems cambiques de la partie Est de la Roumanie

Spécification	HSO + HD 120	LC ♂ HT 228	HS Pioneer 3978	HO 96	HT 228
Concentration en oligoéléments de la semence (en ppm)					
Zn - Témoin	13,3	23,8	18,7	13,3	15,3
ZN - Moyenne trait. foliaires	14,4	24,2	20,5	17,9	17,0
Gain en concentration (% par rapport au témoin)	8	2	10	34	12
Mo - Témoin	0,08	0,13	0,15	0,13	0,0?
Mo - Moyenne trait. foliaires	0,27	0,26	0,27	0,27	0,2?
Gain en concentration (% par rapport au témoin)	237	100	80	107	162

Production de grains obtenue de la semence à teneur optimisée en
oligoéléments (sur agrofond de $N_{100}P_{30}$, kg/ha)

	HSO + HD 120	LC ♂ HT 228	HS Pioneer 3978	HO 96	HT 228
Témoin	7030	3320	8370	7400	7060
Moyenne - effet rémanent trait. foliaires	7920	4450	8810	7860	7930
Gain moyen de production par rapport au Témoin %	13	34	5	6	12

x) On a appliqué 4 engrais foliaires :
    deux engrais produits industriels (F 231 - Craiova et F - Fagaras);
$ZnSO_4.7H_2O$ et
    un concentré en oligoéléments produit par l'Institut de Recherches
pour Pédologie et Agrochimie de Bucarest. (2,5 g $ZnSO_4.7H_2O$;
0,5 g $(NH_4)_6 Mo_7O_{24}.4H_2O$; 1 g H_3BO_3; 0,5 g acide tartrique
- $H_2C_4H_4O_6$; 1,05 g acide citrique $C_6H_5O_7.H_2O$; 0,1 g Na_2 EDTA.
2 H_2O et 1,5 g KOH, à titre de solution appliquée sur les plantes).

Tableau 10

Apport en oligoéléments par les engrais organiques de différentes provenances

Nature des engrais	Zn	Cu	Mn	Fe	B
		g/tonne d'engrais sec			
Fumier (système familial)	60-180	15-62	150-170	2300-20000	20-60
Déchets bovins (système industriel)	10-400	2-307	130-325	60-33000	11-270
Fientes de volailles (sans litière)	158-892	20-680	156-330	360-2630	18-30
Fientes de volailles (avec litière)	155-191	21-155	168-277	600-730	20
Déchets de porcins	3-323	4-700	292	1020	13
Compost de boue de porcins	377-930	141-390	480-810	15000-27000	9-22
Déchets d'ovins	78-126	108-330	170-600	1670-13700	45

Données compilées d'après diverses sources internes et externes (Vintila et allii 1978, Vintila et allii 1981, Berryman et allii, 1966; Coppenet, 1977; Sommers et Sutton 1980; Taiganides P. 1977).

Tableau 11

Exportations en oligoéléments en fonction du niveau des récoltes
(valeurs orientatives pour un niveau optimum de nutrition des plantes)

a) Maïs	4 t grains 6 t tiges	6 t grains 9 t tiges	10 t grains 15 t tiges	15 t grains 22,5 t tiges
Consommations spécifiques x)		Exportations, g/ha		
Fe	100	400	600	1000	1500
Mn	75	300	450	750	1125
Zn	50	200	300	500	750
Cu	10	40	60	100	150
B	10	40	60	100	150
Mo	0,5	2	3	5	7,5

b) Blé	20 q grains 26 q pailles	40 q grains 52 q pailles	80 q grains 104 q pailles	100 q grains 130 q pailles
Consommations spécifiques		Exportations, g/ha		
Fe	160	320	640	1280	1600
Mn	100	200	400	800	1000
Zn	65	130	260	520	650
Cu	11	22	44	88	110
B	9	18	36	72	90
Mo	0,7	1,4	2,8	5,6	7

b) Betterave sucrière	20 t racines 16 t fenilles	50 t racines 40 t fenilles	80 t racines 64 t fenilles	100 t racines 80 t fenilles
Consommations spécifiques		Exportations, g/ha		
Fe	15	300	750	1200	1500
Mn	12	240	600	960	1200
Zn	6	120	300	480	600
Cu	1,5	30	75	120	150
B	9,5	190	475	760	950
Mo	0,1	2	5	8	10

x) En grammes à la tonne de produit principal et la quantité correspondante de produit secondaire (à l'humidité de récolte) (d'après Finek 1982).

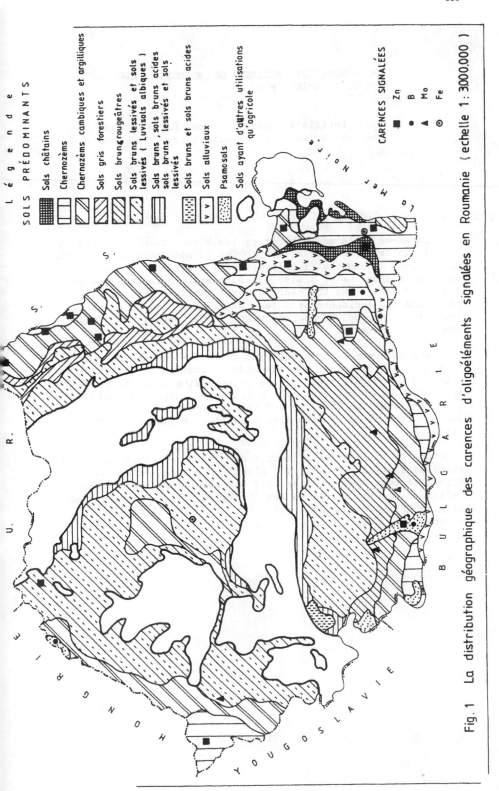

Fig.1 La distribution géographique des carences d'oligoéléments signalées en Roumanie (echelle 1 : 3000.000)

L é g e n d e

S O L S   P R É D O M I N A N T S

- Sols châtains
- Chernozèms
- Chernozèms cambiques et argilliques
- Sols gris forestiers
- Sols bruns rougeâtres
- Sols bruns lessivés et sols lessivés (Luvisols albiques)
- Sols bruns , sols bruns acides sols bruns lessivés et sols lessivés
- Sols bruns et sols bruns acides
- Sols alluviaux
- Psamosols
- Sols ayant d'autres utilisations qu'agricole

CARENCES SIGNALÉES

- Zn
- B
- Mo
- Fe

MICRONUTRIENT PROBLEMS IN INTENSIVE CROP
PRODUCTION IN ROMANIA

Mrs. Leticia Tiganas, Institue for Research
on Soil Science and Agrochemistry, Bucharest

SUMMARY

Numerous field and laboratory tests have been conducted in
Romania in order to assess the problem of micronutrients in crop
production and study various aspects of practical interest. A
considerable volume of analytical data and information has been
collected concerning the micronutrient supply status of agricultural
soils and the incidence of a few nutritional disorders, particularly
involving zinc, molybdenum and boron in various climatic and
management contexts.

For the purpose of evaluating certain micronutrient-caused
nutritional imbalances, use was made in routine agrochemical work of
a series of consolidated indices encompassing chemical properties of
the soil (for example the zinc deficiency, molybdenum and mobile
phosphate reaction indices) and the plant (for example, indices
derived from maize hybrids with essential or induced zinc
deficiency).

On the basis of experimental results obtained in the field and
chemical analysis of soils and plants, practical preventive and
treatment measures were developed for several possible micronutrient
deficiencies with a view to limiting yield losses. For the purpose
of quantifying recommendations relating to fertilization,
mathematical models incorporating the required information are
widely used.

The micronutrient problem is higlighted in intensive agriculture. There is greater demand for the micronutrient reserves of the soil as a result of continuous increases in yields due to the use of certain improved varieties, the large-scale application of chemical NPK fertilizers, irrigation, pesticide use and other means of intensifying crop production. In these circumstances the micronutrient supply from the soil reserves exhibits a clear-cut tendency to fall as yields increase. The nutritional imbalances in plants caused by micronutrients will probably be more frequent in intensified crop production if no micronutrient fertilizer is applied.

In general, it would seem more appropriate to use special micronutrient fertilizers after applying certain management practices such as regular use of suffcent quantities of organic fertilizers, proper crop rotation, raising of the micronutrient content of seeds and cultivation of varieties less susceptible to nutritional imbalances, where such exist. In keeping with this approach, microelement fertilizers are to be used to complement the management measures referred to above.

Seed and leaf applications during the first vegetative stage with special preparations containing micronutrients constitue the most economical means of tackling micronutrient deficiency in agricultural crops.

The limited quantities of micronutrient fertilizer which are currently available should be used first and foremost in seed production, through soil and also leaf applications, in order to obtain high-quality seeds which also have an optimum micronutrient content.

TRACE ELEMENT LEVELS IN SOILS
AND PLANTS IN HUNGARY

Dr. J. Karlinger, Ministry of
Agriculture and Food, Budapest

Hungary is generally poor in natural resources but has a rather
favourable potential of natural resources for agricultural
production. 70% of the total area of the country (6.5 million
hectares) is cultivated. On two-thirds of this area cereals and
maize are produced. Thus significant economic interest is involved
in increasing the yields of crop production.

With the presently available technologies, only plants are able
to transform the heat and light energy of solar radiation free of
charge and with the highest efficiency, into food or industrial raw
material. Consequently, plant production is the only energy
producing process on earth which can be considered as a renewable
energy resource.

In order to increase the yields of crops, new varieties of high
biological potential are continuously required which in turn need a
balanced nutrient supply to be assured in an optimal way. In
addition, pests hindering the development of the biological
potential must be restricted in such a way as to avoid environmental
pollution of any dangerous degree.

On the basis of experiences gained during the last decade it
has been unanimously confirmed that agricultural production cannot
be developed or modernized unless the biological requirements of the
plants are satisfied to an extent which is tolerable from an
economic point of view, not to mention the application of varieties
best suited for production under the given circumstances.

Regarding production technologies particular emphasis should be
given to plant nutrition and plant protection which must be applied
in a harmonious way. If any of the phases of technology are of
lower or higher level that the other, decreasing efficiency and
higher input costs will result instead of the expected enhancing
effect.

Metabolic processes occuring in plant organisms like in other living organisms, are characterized by catalytic processes based on complicated enzymatic and hormonal reactions. In these processes trace elements play a critical role which influences the life cylces in plant organism and are therefore of great importance from the point of view of human and animal nutrition. Under normal conditions plants take nutrients of vital importance from the soil. If any of these nutrients are not available in the soil or if nutrient uptake is inhibited by certain physical or chemical parameters of the soil, the crop will not be able to develop in the region concerned or if growth takes place under artificial circumstances the plant will suffer from diseases caused by nutrient deficiency.

Prior to the initiation of fertilizer application the only method of soil management was the regular application of organic manure. In organic manure, in addition to N, P and K all essential trace elements can be found in optimal rates for the plants. This sort of extensive plant production, however, allowed only low yield averages. A nutrient balance between macro and micro elements was developed where the amount of nutrient intake was equal to the nutrients removed by harvested crops and their by-products.

In Hungary, following the establishment of agricultural co-operatives and state farms between 1960 and 1975, fertilizer consumption increased nearly tenfold and reached 270 kg NPK/ha on the national average. Correspondingly the average yield harvested increased 2.5 fold. Due to the spreading of large-scale production systems and of monocultural production, the plant nutrition pattern which was earlier balanced has become disturbed.

High phosphorous and potassium rates applied inhibited the uptake of certain trace elements and the so called antagonistic phenomenon appeared. After a certain period of time the intensive monocultural crop production regularly calls for micronutrient deficiencies if a reasonable nutrient supply is missing.

The amount of trace elements available to plants is significantly influenced by the pH value of the soil. Some of these nutrients will be water soluble at neutral or slightly alcalic pH values while others, e.g. toxic aluminium, will only be mobilized at an acidic pH value. Acidification of soil causes problems also in Hungary like all over the world. Environmentalists are blaming above all fertilizer application for the acidification and are forgetting the fact that this problem also exists in pine forests which were never treated with fertilizers. Soil acidity is to the greatest extent caused by acid rains. From acidified soil, magnesium and calcium ions are leached into the lower layers of the soil and this adversely affects the micronutrient availability for the crops.

Micronutrient deficiency in the soil due to the various reasons mentioned will hamper the exploitation of the biological potential of the crops and will deteriorate the efficiency of fertilizers.

In pursuit of increasing efficiency of plant production and of fertilization, the Ministry of Agriculture and Food established a nationwide laboratory network for soil and plant analysis in 1976. These laboratories are equipped with automatic analytical instruments which allow uniform analytical methods to be applied. The results obtained are processed by a centralized computer system.

The operation of this system permits an evaluation based on computerized data processing and an extension service on nutrient management needed for achieving the planned yields in the most economical way.

From the data used for developing maps on the macro and micronutrient supplies of soils in Hungary it can be seen that manganese, zinc and copper deficiencies are prevailing on 4, 46 and 9 per cent of the soils of Hungary respectively. These average values are derived from the extreme values measured in the given region. Thus, for example, the rate of areas showing manganese, zinc and copper deficiency varies from 0 to 20, 12 to 87 and 0 to 23 per cent, respectively in the different regions of the country. Data of soil analysis showing the trace element content, however, do not always give accurate information on the occurrence of nutrient deficiency due to factors inhibiting nutrient availability. For this reason widespread plant analysis both for macro and micronutrients is also carried out in each laboratory.

Owing to the Hungarian crop production structure, the majority of plant analysis is made on winter wheat and barley samples. From the data obtained it can be seen that copper and zinc deficiency occur most frequently in winter cereals. On the basis of tests executed at the end of spring 1986, copper, zinc and manganese deficiency were experienced on 46, 24, and 10 per cent, respectively of the total area examined.

During the last few years maize and sunflower were also included in the analysis apart from winter wheat and barley. In the western region of Hungary boron deficiency was experienced in sunflower which was attributed to a long lasting drought period in accordance with the opinion of experts. Maize analysis indicated zinc deficiency in certain regions. In the areas concerned, phosphorous supply of the soils was rather high. Thus it could be assumed that zinc deficiency occurred most probably due to phosphorous and zinc antagonism.

On the basis of the above, an important requirement of harmonized plant nutrition is that, in addition to properly applied NPK nutrients, trace elements should be assured at a rate defined on the basis of plant analysis. To accomplish this requirement, two possibilities are available: (i) improved micronutrient supply and (ii) supplementary fertilization through foliar application. Since extremely high rates are needed for micronutrient fertilization and since the products which meet the requirements are quite expensive, the costs of this technology are rather high. Trace elements applied through cheaper compounds can easily be bound to soil particles and thus become unavailable for the plants.

Another method is offered through foliar application of micronutrients. In Hungary, the use of foliar fertilizers became popular at the beginning of the 1960's. Thus supplementary fertilization can be carried out economically in combination with plant protection operations under practical conditions. The application of foliar fertilizers will be influenced to a great extent in future by the spreading of extension services based on plant analysis as well as by recognizing the fact that micronutrient depletion from the soils calls for a better understanding and for urgent measures of trace element application by the farmers.

TENEUR EN OLIGO-ELEMENTS DES SOLS
ET DES PLANTES EN HONGRIE

M. J. Karlinger, Ministère de
l'agriculture et de l'alimentation,
Budapest

RESUME

En Hongrie, la teneur des plantes et des sols en éléments
nutritifs, y compris les oligo-éléments, est déterminée dans les
laboratoires faisant partie d'un réseau national. Ces laboratoires
qui relèvent de la juridiction du Ministère de l'agriculture et de
l'alimentation utilisent le même matériel et appliquent des méthodes
uniformisées pour faire ces analyses.

La plupart des essais effectués par ces laboratoires,
appartenant aux centres phytosanitaires et agrochimiques des comtés
du pays, portent sur l'évaluation des céréales d'hiver (blé et orge)
et de printemps. La récapitulation des résultats des tests indique
très souvent une carence de cuivre et de zinc dans les échantillons
de céréales d'hiver. Outre le blé d'hiver et l'orge, le maïs et le
tournesol font l'objet de recherches, depuis quelques années, afin
de déterminer les causes des carences.

Dans la partie occidentale de la Hongrie, on a constaté des
carences en bore dans les tournesols, dues probablement à une longue
sécheresse. Dans certaines régions, le maïs présentait une carence
en zinc. Dans les comtés concernés, la teneur du sol en phosphore
est assez élevée, par conséquent la carence en zinc peut avoir été
provoquée par un antagonisme entre le phosphore et le zinc. Les
laboratoires formulent des recommandations relatives à l'apport des
oligo-éléments, soit dans le sol soit en application sur les
feuilles en se fondant sur les résultats des essais. Selon les
experts, ces applications permettent d'augmenter sensiblement les
rendements.

UTILIZATION OF MICRONUTRIENTS
IN TURKISH AGRICULTURE

Dr. N. Ülgen, Director, Soil and
Fertilizer Research Institute, Ankara

I.   INTRODUCTION

In order to fulfill the requirements of rapidly increasing
population of the world, more and more food and fiber must be
produced from each hectare of the world's arable land. It is well
known that, improved varieties of seed, protection of plant from
diseases, pests and weeds, advanced cultivation methods, efficient
use of fertilizers and increase in irrigated areas etc. are all
factors responsible for the general increase in crop production in
recent decades. The use of fertilizers plays a very important role
among these other factors.

In addition to the mineral elements N, P, K, Ca, Mg and S which
are defined as macronutrients or major nutrients, plants also
require other mineral elements which are generally described in the
literature as minor elements, micro elements, micronutrients or
trace elements. These terms are derived from the very small
requirement for these elements for normal plant growth compared with
the macronutrients. Today, about a dozen trace elements, including
iron, zinc, manganese, copper, molibdenum, boron, chloride, sodium
and possibly cobalt are known or suspected to be essential for the
normal growth of plants.

In general all trace elements are toxic to plants and animals
if present in the soil in concentrations appreciably in excess of
the normal or above average. Even though much of the nature of
trace elements functions is known, the application of this knowledge
is not easy. To avoid serious mistakes in this application, much
research and especially extensive experimentation is still needed.

The purpose of this report is to give readers some general
information about the use of fertilizer as well as the need of
micronutrients in Turkish agriculture.

2. AREA, SOILS AND CLIMATE

Turkey has a total area of about 77, 945, 200 hectares of land of which about 26.4 million hectares are suited for cultivation. About 8 million hectares can be irrigated. The soils that are physically suited for cultivation are largely used for cropland, vinyards, orchards, or gardens. In 1985, the total land used in thse forms represented 33.8% of the total area of the country.

Turkey lies between latitudes 36° and 42° North and longtitudes 26° and 45° East. The distance from east to west is about 1530 km, and from north to south, the widest part is about 635 kilometers.

There is a wide range in climatic conditions due to differences in elevation and the location of mountain ranges which act as barriers to moisture bearing winds. The greater portion of the country is subhumid and lack of moisture is one of the principal limiting factors in crop production. Areas receiving adquate summer rainfall for warm season crops are rather small. Vegetation varies according to climate. The original vegetation consists of forest of pine, fir, bearch and oak in the more humid sections and a tin cover of grass and scattered shrubs in the semiarid interior.

The dominant soils in Central Anatolia are of Reddish Brown and Brown Soil groups. These soils are used to a great majority for small grains, mainly wheat and barley. These are dominantly finely textured, medium to shallow depth calcareous soils with nearly level to rolling topography.

The Alluvial soils comprise a large and important group of arable soils in Turkey. These are mostly fine to moderately fine textured calcareous soils with more favourable moisture relations than that of other soil groups. Almost all kinds of crops and a wide variety of vegetables and fruits are grown on thse soils.

In Thrace and south of Marmara Sea there are large areas of undulating to rolling dark clayey calcarous soils of Grumusol and Rendzina groups with intermixed areas of lighter-colored non calcareous, medium textured Noncalcic Brown soil group (Oakes, 1956).

Some of the physical and chemical characteristics of Turkish Soils are summarized in Table 1. It can be seen from Table 1 that the great majority of Turkish soils (85%) are of loam and clay loam in texture. Most of them are calcareous soils, some containing up to 60-70 per cent of calcium carbonate. In a relatively small area along the Eastern Black Sea coast, soils are acid in reaction. Most soils in Turkey are relatively low in organic matter, nitrogen and available phosphorus, but sufficient or high in potassium content (Ulgen 1984).

Table 1

Some Physical and Chemical Characteristics of
Turkish Soils (Ulgen, 1984)

aturation %	<30 Sandy	30-50 loam	50-70 Clay loam	70-110 Clay	110 heavy clay
ercent Distribution	3.4	37.9	47.9	10.4	0.4

H (Saturation paste)	4.0-4.9	5.0-5.9	6.0-6.9	7.0-7.9	8.0-8.9
ercent Distribution	0.9	4.5	13.4	76.5	4.7

ime content %	< 1	1-5	5-15	15-25	>25
ercent Distribution	22.0	20.4	26.9	18.8	11.9

rganic Matter %	<1 very low	1-2 low	2-3 medium	3-4 sufficient	>4 high
ercent Distribution	19.2	49.8	22.4	5.6	3.0

vailable phosphorus OLSEN, kg P_2O_5/da)	<3 very low	3-6 low	6-9 medium	9-15 sufficient	>15 very high
ercent Distribution	33.3	32.8	18.0	8.7	7.2

vailable Potassium Ammonium Acatate kg K_2O/da)	> 20 Low	20-30 medium	30-60 sufficient	60-100 High	>100 Very high
ercent Distribution	1.8	2.3	13.0	27.2	55.7

3. FERTILIZER CONSUMPTION IN TURKEY

Fertilizer consumption in Turkey for the period of 1980-1985 is given in Table 2. As seen in Table 2, the plant nutrients being used in Turkey are mainly nitrogen and phosphorus. The use of nitrogen increased from about 640.000 tons to one million tons, while phosphorus increased from about 480.000 to 575.000 tons in the period of 1980-1984. The ratio of N to P_2O_5 was 1.3 in 1980 and it reached 1.7 in 1984. It means that the rate of increase in the use of nitrogen in this period of time is greater than the rate of use of phosphorus. In the same period, the use of potassium remained almost the same.

Table 2

Fertilizer consumption in Turkey (1980-1985)

Years	Plant Nutrients (1000 Tons)			
	Nitrogen (N)	Phosphorus (P_2O_5	Potassium (K_2O)	Total
1980	638	483	44	1165
1981	776	495	38	1309
1982	847	570	33	1450
1983	991	618	25	1634
1984	998	575	31	1604
1985	920	476	34	1430

Source: Ministry of Agriculture, Forestry and Village Affairs.

4. FERTILIZER DEMAND THROUGH 1987-1993

Fertilizer needs projected for Turkish agriculture in the period of 1987-1993 is given in Table 3. As seen in this Table Turkey will be using 1.400.000 tons of N, 1.000.000 tons of P_2O_5 and 30.000 tons of K_2O in 1993. A total of 2.5 million tons of plant nutrient is assumed to be used at the end of the sixth Five Year Plan. After realization of that amount of fertilizer use, in 1993 about 100 kg plant nutrients will be used for one hectare of arable land in Turkey.

Table 3

Demand of Fertilizers in Turkey

(Plant nutrients in 1000 tons 1987-1993)

Years	Nitrogen (N)	Phosphorus (P_2O_5	Potassium (K_2O)
1987	1230	960	23
1988	1260	990	24
1989	1300	1020	26
1990	1330	1050	27
1991	1370	1080	29
1992	1400	1100	30
1993	1430	1140	32

5. MICRONUTRIENT DEFICIENCY OF THE SOILS IN TURKEY

This presentation shall be limited to summarize some of the investigation that reveal micronutrient deficiencies of Turkish soils.

For a long time it was felt that under the present farming systems and the fertilization programmes, the micronutrient status of the soils was adequate and the micronutrient deficiencies may not be a serious problem. However, indications from many countries show that micronutrient problems are becoming more and more frequent.

Under the direction of Sillanpää (1982) a research project called "Trace Element Study" started in 1974 under the FAO/Finland Cooperative Programme which involved a worldwide study on micronutrients, in cooperation with 30 countries.

The purpose of this study was to obtain more information on the problems of various micronutrients under different soil and climatic conditions in participating countries.

Turkey is one of the countries that participated in this study. 250 wheat-soil and 50 maize-soil samples were collected from agriculturally important areas of Turkey and were analyzed for macro and micro nutrients. The majority of the soil groups that were taken into the study were classified by FAO/UNESCO as the following soil units: Fluvisols, Luvisols, kastanozems, Vertisols, Lithosols, Rhegosols and Xerosols.

According to the analyses of original plant samples (wheat and maize) from the field and the respective soils for diagnosing the micronutrient status of soils the investigators summarized the results as follows:

1. Zinc contents of Turkish soils and plants are low especially in the samples which are collected from Central and Eastern Anatolia. About 20% of the samples from the Black Sea, Marmara and Aegean regions fell into the low zones. Zinc contents of Turkish soils and plants are amongst the very lowest recorded in this study. The most evident micronutrient disorders in Turkey are those due to deficiency of Zn.

2. Iron contents of Turkish soils and plants are low compared with most of the other countries. It seems that in many places crops sensitive to iron deficiency would respond to iron fertilization. Almost 90% of the low iron values occur in Central Anatolia, mostly in the Province of Konya.

The level of iron in Turkish soils is also low as that of Zn.

3. The level of Mn is also low. However, in spite of the generally low Mn content, only about two percent of the Turkish samples fall into the lowest Mn zone, and none of these indicates any severe shortage of available Mn.

4. The Turkish national average of soil and plant Cu contents correspond closely to the international respective mean values in this study. Extremely low or high Cu contents were not recorded. The present analytical data suggest that problems due to shortage or excess of Cu are unlikely in Turkey.

5. Boron contents of Turkish soils and plants are high in comparison with the other countries. In spite of wide variations, 84 per cent of the plant and soil B values are within the "normal" B range, and only two percent of these, fall into the two low B zones. The lowest B values occur in the Black Sea, the Aegean and Marmara regions. The majority of the high B values are found in Central Anatolia. In many cases the practice of irrigation is partly responsible for the high B values.

Analytical data obtained from Turkish samples indicated that B deficiency occurs in Turkey but is not widespread. B values are usually normal, but at several locations attention should be paid to possibe disorders due to an excess of available B caused either by naturally high B contents of soils or induced by high-B irrigation waters.

6. Only a few soil and plant samples indicate a shortage of available Mo. Several relatively high, but not extreme, Mo values were recorded from the samples. In general, the Mo values are normal in Turkish soils.

Ulgen and Aksu (1962) studied the effect of micronutrients on alfalfa yield in greenhouse experiments conducted in two years on calcareous soils taken from the Atatürk Farm in Ankara. Micronutrients like Zn, Cu, and Mn with NP fertilizers were used either alone or in combinations. After harvesting five cuttings of alfalfa the yields were evaluated. The results indicated that the minor elements used in this investigation produced a significant increase in yields of alfalfa.

Ulgen and Bellik (1962) worked with the pot culture technique utilizing lettuce as a test plant to study the needs of phosphorus and micronutrients of Cu and Zn, on calcareous soils of the Atatürk Farm in Ankara. Evaluating the data obtained from the experiment, they concluded that yield increases were significant with phosphorus, but zinc and copper in addition to phosphorus resulted in significantly higher yields.

To study the needs of micronutrients on Central Anatolian soils, Ulgen et al. (1965) conducted a field experiment in the Gözlü State Farm with alfalfa using Cu, Zn and Co which were found deficient in the previous greenhouse experiment. The above mentioned micronutrients alone and their combinations were used in a split plot design. Data obtained from the experiment indicated that Cu, Zn and Co used either alone or in combination, increased the alfalfa yield significantly. They continued the experiment over the next two years by applying micronutrients and phosphorus in the spring of each year. The results of the next two years' experiments were almost identical. There were some indications that the yields of alfalfa increased by application of minor elements used in the study though it was not statistically significant. This result applies only for the soil tested.

The widespread appearance of characteristic leaf patterns in orange orchards in all Mediterranean citrus growing regions of Turkey were studied by Ozbek (1970). Soil properties of the sampling units, including their macro and micro nutrient status were determined. Leaf samples taken from fruiting terminals were analyzed for micronutrient contents. The results of this study indicate that the characteristic leaf patterns were due largely to zinc, and to a very minor extent to iron deficiency. The degree of zinc deficiency varied between moderate to acute, depending particularly on soil properties and orange varieties. Zinc deficiency was more acute on soils with lower pH, lower $CaCO_3$ and lighter texture. Iron deficiency appeared only in one region which has soils of quite high pH, high $CaCO_3$ and moderately heavy texture.

In order to find out suitable methods of application to correct iron deficiency; Ozbek and Danisman (1973) conducted field experiments on orange varieties grown in Mediterranian coastal area of Turkey. Soil application of Fe-chelates as well as stem injection of inorganic iron solutions were tried. Leaf and fruit analyses were performed before and after iron applications. According to the data obtained from the experiments, the investigators concluded that to correct Fe-deficiency and to improve both yield and fruit quality of oranges, soil application of Fe-chelate can be recommended.

To find the need of micronutrients for apple trees showing the symptoms of chlorosis Ulgen et al. (1973) conducted twi field experiments in two countries (Beypazari and Cubuk) of Ankara. Micro nutrients of Zn, Fe, Mn and Cu were sprayed in he chelate form in 15 days intervals. The results obtained from the experiments showed that all treatments containing Fe were effective on chlorosis.

Ulgen et al. (1973) conducted three field experiments to determine the effect of various micronutrients on chlorosis. Cu, Zn, Mn and Fe were sprayed in chelate form in June, July and August. The results indicated that in general, all treatments containing Fe were effective to reduce the chlorosis of apple, pear and quince trees growing in the vicinity of Ankara.

Field experiments and laboratory tests were made by Ozbek et al (1977) to diagnose and correct macro and micro nutrient deficiencies in important grapefruit varieties grown in the Mediterranean region of Turkey. Soil properties, including their macro and micro nutrient contents, were determined. In order to diagnose the deficiency of macro and micro nutrients visual symptoms were supplemented by leaf and soil analyses. The experimental data indicated that there is no macro nutrient deficiency problem, whereas Fe, Zn and Mn were deficient in these grapefruit growing areas, and application of Fe chelate and inorganic salts of Zn and Mn were found effective in solving the problems of deficiencies.

Ozbek et al. (1977) conducted field experiments to diagnose and also to correct micronutrient deficiency in important lemon varieties grown in the Mediterranean region of Turkey. Soil properties of the experimental areas, including their macro and micronutrient contents were determined. Leaf samples were also taken periodically for chemical analyses. In evaluating the results obtained from the experiments, the investigators concluded that in Alata and Antalya iron deficiency appeared on the soils having high pH and CaCO$_3$ contents. The degree of iron deficiency varied among the lemon varieties. Characteristics leaf patterns were due largely to iron and to a lesser extent to zinc deficiency. By various Zn and Mn-treatments much higher yields were obtained as compared to control, whereas there were no significant effects of these treatments on fruit quality including vitamin C.

Kacar et al. (1979) conducted a study to find out the effects of micronutrients on the yield of tea plants and to determine the micronutrient contents of tea plants and tea growing soils in Turkey. Thirty representative tea plantations were selected in the surroundings of Rize (East of Black Sea Coast) and green tea leaf samples as well as composite soil samples were collected from the same tea plantations. In the field experiments NPK fertilizers were mixed with the soils and different micronutrients were sprayed on leaves three times in each growing season.

Soil and plant samples were analyzed for macronutrients as well as micronutrients such as Cu, Zn, Fe, Mn Mo and B. Evaluating the data obtained from the experiments they concluded that a significant increase in the yield of tea leaves, as compared to control, was obtained with NPK and through the application of micronutrients by spraying after NPK. A maximum increase in yield, as compared to NPK, was obtained with the application of B, Zn, Cu and Mo together. Nevertheless, they also concluded that according to the critical values reported by different researchers there is no important deficiency problems for microelements in the tea plantations of farmers for the time being. Similar results were also obtained for the tea growing soils.

The macro and micro nutrient status of the soils in Nevsehir province was studied by Aksoy (1979) to find the nutritional problems of potatoes grown in this region. Soil samples taken from 22 locations within the province of Nevsehir were analyzed for macronutrients as well as micronutrients such as Fe, Zn, Cu, Mn. Chemical analyses indicated that soils of experimental sites are sufficient in Fe, except some locations, mainly deficient and moderately deficient in Zn, and sufficient or/and high in Mn. Experiments conducted in Zn deficient soils indicated that yields of potatoes were increased significantly by application of 10 kg $AnSO_4$ per decare.

In a greenhouse experiment Aydeniz et al. (1978) found that application of 5 kg/ha of zinc sulfate to rice grown on soils taken from Corum, Ankara, Samsun, Balikesir, Edirne and Bursa increased the yield of dry matter by between 3.46 and 10.81 %.

Ergene et al. (1980) studied the available Fe, Mn, Zn and Cu status of the soils of the Malatya province which has a high fruit production potential. 38 soil samples from highly calcareous land and 61 plant samples from different fruit trees were analyzed for macro and micro nutrients. According to the data obtained from the experiment, the investigators concluded that Zn and Mn are deficient in soils. Therefore, in addition to macronutrients Zn and Mn should be added to these soils to increase the yields of different fruits growing in this region.

In order to diagnose and correct he efficiency of micronutrients of citrus trees in Eagian and Mediterranean regions of Turkey Alemdar et al. (1980) conducted 12 field experiments on tangerine, orange and lemon in different provinces of these regions. Soil and plant samples taken from the experiments were analysed for macro and micro nutrients. Data obtained from the experiments indicated that, although Zn was superior, Zn and Mn were both effective for recovering the chlorosis of citrus in Köycegiz, Ula, Kuyucak and Silifke. In Izmir the effectiveness of Fe was greater than of Zn.

Effect of Zn fertilization in addition to NP fertilizer on the yield and quality of rice were studied by Karaçal and Teceren (1983) at Zn deficient locations in Central Anatolia and Thrace in Turkey. Field experiments conducted in Corum, Ankara and Edirne provinces were continued for two years in farmer's fields. Available Zn, Cu, Mn and Fe in the soils of experimental sites as well as Zn, Cu, Mn and Fe contents of the plant samples were determined. Available Zn content of the soils, especially soils from Osmancik-Corum were found to be in the critical range. Sillanpää (1982) also reported this area as deficient in available Zn. Other micronutrients such as Fe, Mn and Cu were found sufficient in these soils. It was found that in Zn deficient soils of the paddy rice area of Osmancik, application of 3-6 kg of $ZnSO_4$ to the soil in addition to NP fertilizer increased the yield significantly. In the other two locations Edirne and Nallihan-Ankara they recommended to spray the zinc sulfate twice to the paddy field insteadof soil application.

In order to estimate the ability of plants to absorb soil zinc Yalçin (1983) conducted a greenhouse experiment by using soil samples taken from different counties of the Konya plain in Central Anatolia. Available Zn content of the soil samples analyzed by different extractants indicated that soils of two locations were deficient in Zn and the remainder was in the critical range. By evaluating the data, the investigator concluded that application of increasing rátes of Zn to these soils increased the yieds of six different crops significantly, with maize showing the highest response whereas the response of wheat was the lowest.

Türkoglu (1974) has found that apple trees growing in several countries of the Konya and Nigde provinces in Central Anatolian Plateau on highly calcareous soils showed chlorosis due to iron deficiency. Application of chelate materials as well as of inorganic iron sulfate to the soils and spraying to the leaves corrected the chlorotic symptoms of apple trees. Frequency of chlorosis was higher on soils containing moe than 20% of lime.

Kurucu (1986) reported a study which was conducted by Soil Fertility Staff of the Soil and Fertilizer Research Institute of Ankara to determine the effect of various micronutrient fertilizers on correcting the symptoms of Cholorosis of apple and peach trees. Two-year field experiments carried out in the Nevsehir, Konya, Usak and Bursa provinces on moderate to highly calcareous soils with low and nearly sufficient amounts of Fe and Zn indicated that iron chelates as well as iron sulfate applied to the soils were effective to correct the chlorosis of apple and peach trees. They also concluded that since the chlorosis were due to Fe deficiency on these calcareous soils, other micro nutrients such as Zn and Mn, either in chelate or inorganic form were not effective.

Kurucu et al. (1986) carried out field experiments in order to find out the micronutrient deficiencies in chloratic hazelnuts grown on the Black Sea coast. NPK fertilizers as well as trace elements like Fe, Mn and Zn were applied either to the soil or sprayed to the leaves. Leaf samples taken from the experiments were analyzed for NPK and Fe, Zn, Mn. Evaluating the data obtained from the experiments, investigators indicated that NPK fertilizers with 200-300 gr of iron chelate or iron sulfate with farmyard manure applied around the trees reduced the chlorosis. They also found that spraying of iron chelate three times with 20-25 days intervals were effective as well.

A field experiment, conducted in micronutrient deficient locations which were indicated by earlier work of silanpää (1974) is reported by Sungur (1986). Field trials were carried out in 13 provinces of Turkey. In this investigation micro nutrients like B, Cu, Fe, Mn, Mo and Zn together with macro elements of NPK, Mg and S were used to grow wheat, barley, sunflower, maize and rice as test plants in 19 field experiments. Soil and plant samples were taken for chemical analysis, including available Fe, Zn Cu and Mn in the soils of experimental sites. From the data obtained from the trials, it can be concluded that remarkable yield increases were obtained on paddy rice by application of zinc in the Aydin-Söke experiment, while the other micronutrients indicated in significant effects on the yields.

An evaluation of the micronutrient status of some olive plantations around Ayvalik County in the Eagian Region reported by Kovanci et al. (1986) indicated that there are some places in which soils are deficient in Fe, Zn and Mn. They also concluded that Fe contents of the olive leaves were lower than the critical levels, whereas Zn and Mn contents of the leaves were found to be within the normal range.

6. SUMMARY AND CONCLUSIONS

Turkey has a total area of about 77.945.200 hectares of land of which about 26.4 million hectares of land are suited for cultivation. About 8 million hectares can be irrigated.

The dominant soils in Central Anatolia are of Reddish Brown and Brown Soil Groups. The alluvial soils comprise the large and important group of arable soils in Turkey. In Thrace and south of Marmara Sea large areas covered with Great Soil Groups of Grumusols and Rendzinas.

The majority of Turkish soils are calcareous, loam and clay loam in texture. In a relatively small area along the eastern Black Sea coast, soils are acid in reaction. Most soils in Turkey are relatively low in organic matter, nitrogen and available phosphorus, but sufficient or high in potassium content.

Fertilizer consumption in Turkey is increasing rapidly. The plant nutrients used are mainly nitrogen and phosphorus. In 1985 about one million tons of nitrogen, 600.000 tons of phosphrus and 30.000 tons of potassium were used.

It is projected that 1993 a total of 2.5 million tons of plant nutrients, of which about 1.4 million tons of N, 1.1 million tons of P_2O_5 and 30.000 tons of K_2O, will be used. After realization of that amount of fertilizer application about 100 kg of plant nutrients will be used for one hectare of arable land in Turkey in 1993.

In this presentation, investigations revealing micronutrient deficiencies of Turkish soils are summarized. According to the results of the researchers the following conclusions can be drawn.

1. Micronutrient deficiencies in the soils of Turkey are becoming more and more pronounced and might become a serious problem in the agriculture of the country.

2. In order to solve the problems, micronutrient deficient locations of the country have to be identified and the missing micronutrients should be given to these soils or plants to produce higher yields with better quality.

3. The main micronutrients that are deficient in the soils of Turkey are zinc and iron.

4. In some locations manganese and copper were found to be deficient.

5. Especially in Central Anatolia and in Mediterranean regions lime induced chlorosis is widespread on highly calcareous soils, and greater response to Fe and Zn application can be obtained.

6. Deficiency of Zn in citrus growing areas in the southern Anatolia region is quite large and the use of Zn fertilization is common practice.

7. Organic and inorganic forms of micronutrients can be used satisfactorily to correct the micronutrient deficiencies.

8. There are about a dozen private companies that are dealing with micronutrients in Turkey. It is estimated that about 30.000 kg of micronutrients in chelate form are being used in the country and it could be possible that about 500.000 kg of materials containing micronutrients might be used in the near future.

9. In order to establish a sound micronutrient fertilization programme, the micronutrient status of the soils as well as the response studies of different crops to various micronutrient deficient soils in the country.

10. To achieve a better understanding of the micronutrients and their application in agriculture much further research and experimentation is needed.

156

REFERENCES

1. Aksoy, T. 1979. Nevsehir yöresinde yetistirilen patateslerin
    beslenme sorunlari ve giderilmesi. TUBITAK, TOAG-274,
    Ankara.

2. Alemdar, N., Kurucu, N., Sungur, M., Yürür, H., Eyüboglu, F.,
    Gedikoglu, I., Ors, G. 1980. Ege ve Akedeniz bölgesi
    Turuncgil alanlarinin mikrobesin maddeleri statüsü ve
    mikrobesin maddeleri noksanliklarinin teshisi ile bu
    noksanliklarin giderilmesinde uygulanacak yöntemlerin
    saptanmasi. Toprak ve Gübre Aras.Enst.1979-1980 Yili
    Arastirma Raporu, Sayfa 20-30, Ankara.

3. Aydeniz, A., Danisman, S. 1978. The Response of Zinc on
    Calcareous soil under Flooded condition. IAEA Research
    Co-ordination Meeting on Isotope Aided Micronutrient
    studies in rice production with special reference to zinc
    deficiency. Bogor, Indonesia.

4. Ergene, A., Topbas, M.T., Aydemir, O., Karakaplan, S. 1980.
    Malatya ve yöresi topraklarinin bazi mikrobesin elementleri
    durumlari ile bölgede yetistirilen önemli meyve çestleri
    arasindaki iliskilerin arastirilmasi. TUBITAK, TOAG No.
    362, Ankara.

5. Kacar, B., Przemeck, E., Ozgümüs, A., Turan, C., Katkat, V. and
    Kayikcioglu, i. 1979. Türkiye'de çay tarimi yapilan
    topraklarin ve çay bitkisinin midroelment gereksimmeleri
    üzerinde bir arastirma. TUBITAK, TOAG-321. Ankara.

6. Karaçal, I. and Teceren, M.1983. Celtik tariminda azot ve
    fosfor ile birlikte uygulanan çinko gübrelemesininürün
    miktari ve kalitesine etkisi. TUBITAK, TOAG-442, Ankara.

7. Kovanci, I. and Eryüce, N.1986. Ayvalik zeytinliklerinde
    yaprak ve topraklarin Demir, Cinko, Mangan içerikleri ve
    bitkideki degisimleri. Toprak Ilmi Dernegi 9. Bilimsel
    Toplanti Tebligleri, Yayin No.4, Sayfa 35-1:35-25, Ankara.

8. Kurucu, N. 1986. Iç Anadolu ve Marmara Bölgelerinde mikro besin
    maddeleri kapsayan gübrelerin elma ve seftali agaçlarinda
    etkenlik derecelerinin saptammasi. Tarim Orman ve Köyisleri
    Bakanligi, Köy Hizmetleri Genel Md., Toprak ve Gübre
    Arastirma Enst.Müd.Yayinlari, No.117/R.55, Ankara.

9. Kurucu, N. 1986. Ordu ili çevresinde yaygin sar ma ârâzi gösteren findik alanlarinin makro ve mikro besin maddeleri bakimindan bitki beslenme sorunlarinin teshisi ve giderilmesi. Tarim Orman ve köyisleri Bakanligi, Köy Hizmetleri Genel Müd., Toprak ve Gübre Arastirma Enst.Müd.Yayinlari No.136/R-61, Ankara.

10. Oakes, H.1956. The soils of Turkey. Ministry of Agriculture, Soil Conservation and Farm Irrigation Division No.1, Ankara.

11. Ozbek, N.1970. Akdeniz turunçgiller bölgesinde portakal bahçelerinde ortaya çikan mikro besin maddeleri noksanliklarinin teshisi. A.U.Zirat Fakültesi Yilligi, 19.Yil, Ankara.

12. Ozbek, N. and Danisman, S.1973. Akdeniz Turunçgil Bölgelerinde yetistirilen belli basli portakal çesitlerinde ortaya çikan demir noksanliginin giderilmesinde uygulanacak metotlar üzerinde bir arastirma. Ankara Universitesi Ziraat Fakültesi Yilligi, S.530-560, Ankara.

13. Ozbek, N., Ozsan, M., Danisman, S. and Tuzcu, O.1977. Akdeniz Bölgesinde yetistirilen önemli Altintop çesitlerinde makro ve mikro besin maddeleri noksanliklarinin teshis ve giderilmesi. TUBITAK, TOAG-236. Ankara.

14. Ozbek, N., Ozsan, M., Danisman, S.1977. Akdeniz Bölgesinde yetistirilen önemli limon çesitlerinde görülen mikro besin maddeleri noksanliklarinin teshis ve giderilmesi. TUBITAK, TOAG-144, Ankara.

15. Sillanpää, M.1982. Micronutrients and the nutrient status of soils: a global study. FAOSoils Bulletin No.48, Rome.

16. Sungur, M.1986. Mikrobesin maddeleri ile gübrelemenin ülkemizin degisik yörelerinde yetistirilen bazi kültür bitkilerinin verimine etkileri. Tarim Orman ve Köyisleri Bakanligi, Köy Hizmetleri Genel Md., Toprak ve Gübre Arastirma Enst.Müd.Yayinlari, No.135/R-60. Ankara.

17. Türkoglu, K. 1974. Orta Anadolu Bölgesinde elma plantasyonlarinda görülen kloroz arazinin toprak tipleri ve elma çesitleri ile iliskisi ve en uygn tedavi metodu üzerinde arastirmalar. TUBITAK, TOAG-86, Ankara.

18. Ulgen, N. and Yurtsever, N.1984. Türkiye Gübre ve Gübreleme Rehberi. Tarim Orman veKöyisleri Bakanligi, TOPRAKSU Genel Md., Arastirma Dairesi No.47-8, Ankara.

158

19. Ulgen, N. and Aksu, S.1962. Minör element arastirmalari. Toprak ve gübre Arastirma Enstitüsü 1959-1960 Yili Arastirma Raporu, Sayfa 120-137, Ankara.

20. Ulgen, N. and Bellik, S.1962. Fosfor ihtiyaci arastirmalari. Toprak ve Gübre Arastirma Enstitüsü 1959-1960 Yili Arastirma Raporu, Sayfa 114-119. Ankara.

21. Ulgen, N., Uygun, S., Tezer, G.1965. Minör element ihtiyaci arastirmalari. Tarim Orman ve Köyisleri Bakanligi, Toprak ve Gübre Arastirma Enstitüsü 1963-1965 Yillari Arastirma Raporu, Sayfa 210-213, Ankara.

22. Ulgen, N., Aksu, S. and Selimoglu, F.1973. Meyve agaçlarinda iz element ihtiyaci arastirmalari. Toprak ve Gübre Arastirma Enstitüsü 1969-1971 Arastirma Raporu, Sayfa 82-85. Ankara.

23. Ulgen, N., Uygun, S., Aksu, S., Isik, H., Selimoglu, F. and Kurucu, N. 1973. Meyve agaçlarinda iz element ihtiyaci arastirmalari. Toprak ve Gübre Arastirma Enstitüsü 1969-1971 Arastirma Raporu, Sayfa 86-90, Ankara.

24. Yalçin, S.R.1983. Degisik Kültür bitkilerinin çinkodan yararlanma yeteneklerinin izotop teknigi ile belirlemesi üzerinde bir arastirma. TUBITAK, TOAG-446. Ankara.

UTILISATION DES MICRONUTRIMENTS DANS L'AGRICULTURE EN TURQUIE

Dr. N.Ülgen, Directeur,
Institut de recherche des sols
et des engrais
Ankara, Turquie

RESUME

La Turquie couvre une surface totale de 77 945 200 hectares, dont environ 26,4 millions sont cultivables. Quelques 8 millions d'ha peuvent être irrigués.

Les sols dominants en Anatolie centrale appartiennent aux groupes des sols bruns rougeâtres et des sols bruns. Les terrains alluviaux constituent en Turquie la partie la plus importante des terres arables. En Thrace et au sud de la Mer de Marmara, de vastes zones sont couvertes de sols appartenant aux grands groupes des grumusols et des rendzines.

La majorité des sols turcs sont de texture calcaire, limoneuse et argilo-limoneuse. Une région assez petite, le long de la côte orientale de la Mer Noire, présente des sols de réaction acide. La plupart des sols en Turquie sont relativement pauvres en matière organique, en azote et en phosphore disponible, mais suffisamment ou fortement pourvus en potassium.

La consommation d'engrais s'accroît rapidement en Turquie. Les substances nutritives actuellement utilisées sont surtout l'azote et le phosphore. En 1985, les quantités ont été d'environ un million de tonnes d'azote, 600 000 tonnes de phosphore et 30 000 tonnes de potassium.

Selon des projections faites pour 1993, au total 2,5 millions de tonnes de substances nutritives seront employées, dont environ 1,4 million de N, 1,1 million de P_2P_5 et 30 000 tonnes de k_2O. Lorsque ces volumes seront atteints, en 1993 on utilisera en Turquie autour de 100 kg d'éléments nutritifs par ha de terre arable.

Dans cet exposé sont résumées les investigations montrant les déficiences des sols turcs en micronutriments. Les résultats de ces recherches conduisent aux conclusions suivantes :

1.   Les déficiences en micronutriments des sols de Turquie deviennnent de plus en plus prononcées et pourraient poser un problème sérieux pour l'agriculture du pays.

2.   Afin de résoudre les problèmes, les zones déficientes doivent être identifiées et les micronutriments manquants doivent être fournis aux sols ou aux plantes affectées pour obtenir des rendements plus élevés de meilleure qualité.

3.   Les principaux micronutriments déficitaires sont le zinc et le fer.

4.   Dans certaines zones, on a trouvé des manques en manganèse et en cuivre.

5.   Particulièrement en Anatolie Centrale et dans les régions méditerranéennes, la chlorose due à la chaux est répandue sur les sols hautement calcaires, et une plus forte réponse à l'application de Fe et de Zu peut être obtenue.

6.   La déficience en Zu dans la région de culture des agrumes du sud de l'Anatolie est assez forte et la fertilisation en Zu est une pratique courante.

7.   Les formes organiques et inorganiques de micronutriments peuvent être utilisées de façon satisfaisante pour corriger les déficiences.

8.   Il existe environ une douzaines de compagnies privées qui s'occupent de micronutriments en Turquie. On estime à environ 30 000 kg la quantité de micronutriments sous forme chélate qui est utilisée dans le pays et il se pourrait que 500 000 kg de produits contenant des micronutriments le soit dans un proche avenir.

9.   Afin d'établir un bon programme de fertilisation en micronutriments, les bilans des sols ainsi que les études des réponses des différentes récoltes aux divers micronutriments doivent être menés à terme pour les sols déficitaires du pays.

10.   En vue d'arriver à une meilleure compréhension des micronutriments et de leur application dans l'agriculture, beaucoup de travaux de recherche et d'expérimentation sont encore nécessaires.

ZINC APPLICATION TO CEREALS, POTATOES
AND RED CLOVER ON A HEAVILY LIMED
ZINC-DEFICIENT SOIL IN NORWAY

Dr. I. Aasen, Department of Soil
Fertility and Management of the
Agricultural University, Aas

INTRODUCTION

Micronutrient deficiencies in farm crops, vegetables, and fruit
are well known in Norway. Copper deficiency is mostly found on peat
soils and sandy soils, while B deficiency also occurs on clay
soils. Manganese deficiency is always connected with high soil pH
and loose soil structure. Iron deficiency is known on acid peat
soils only. Application of B, Cu, and Mn in order to correct the
deficiencies have been used for more than forty years. Application
of Mo as a plant nutrient started in Norway about 35 years ago.
(ODELIEN 1967).

The necessity of B fertilization to prevent frost damage and
shot dieback in Norway spruce and Scotch pine on afforested open
ombrogenous peatlands has been established recently (BRAEKKE 1979).

Zinc deficiency in cultivated plants was hardly known in Norway
up to 1970, only a few cases in orchards were reported. However, in
districts with low soil Zn content, Zn deficiency was known in dairy
cattle.

During the last 15 years Zn deficiency in plants has occurred
more frequently, mostly on heavily limed soils. The deficiency has
been found on vegetables, strawberries, fruit crops, cereals and
potatoes. In pot experiments with soils low in plant available Zn,
the deficiency has also been found in sweet corn, carrots, forage
rape, grasses and red clover (AASEN 1978).

Zinc deficiency is in most cases linked to heavily limed or
calcareous soils, and is found both on sandy soils, clay soils, and
moraine soils provided the soil pH is high. Manganese deficiency
often occurs together with Zn deficiency.

The solubility of soil Zn drops drastically when pH increases
up to 7 or higher (LINDSAY 1972). Therefore, application of Zn as a
spray solution on the leaves should be more effective than soil
application.

In order to test the effect of foliar applied Zn on different crops and at different growth stages, a field experiment was conducted over several years. Results from the first seven years of the experiment are summarized in the following report. A few analytical data from two additional years will also be presented.

Materials and methods

The soil was a Zn deficient, heavily limed sandy loam with the following chemical characteristics: 2.2% organic C; pH 7.5 (soil: H_2 = 1:2.5); 1.6 ppm of 0.2 N HCL−extractable Zn (ELLIS, DAVIS and THURLOW 1964); and 11 me. titratable alkalinity/100 g airdry soil (NELSONm, BOAWN and VIET 1959).

Test crops from the first to the seventh years were: Barley, oats, wheat, potatoes (2 years), barley with undersown red clover, and red clover. All the grain crops were spring cereals.

Plant samples for Zn determination were collected every year. Samples from the first three years were digested in HNO_3 and $HClO_4$ mixtures, while for the following years the plant samples were dry ashed and the ash treated on a hot plate with a mixture of HNO_3 and HCl, and finally dissolved in diluted HCl. The Zn concentration in the plant and soil samples were determined by atomic absorption spectrophotometry.

The experimental design was a randomized block experiment with four replicates. The plot size was 21 m^2.

A basic dressing of N, P, and K was given in the spring to all crops, except for the red clover which got no N fertilization.

The experimental treatments for cereals were: a) Control. b) Foliar application of 1% solution of $ZnSO_4.H_2O$, 250 1/ha, 1, 2 and 3 weeks after emergence of the seedlings, and their combinations, i.e. 1+2, 1+3, 2+3, and 1+2+3 weeks after emergence. The potato crop was sprayed the second year only, and at a later growth stage, i.e. 4, 5, and 6 weeks after emergence of the plants, and their combinations similar to the cereal treatments. In order to ensure good contact with the leaves, a surfactant was added to the spray solution.

The red clover was not sprayed with Zn solution, only the residual effect of Zn applications in previous years was measured.

Results and discussion

Grain yield

Table 1 shows the grain dry matter yield. In spite of low yield levels due to reduced tillage, especially in 1977 and 1979, there was a significant yield response to Zn application (p < 0.001). Except for the Zn application one week after emergence, which apparently was too early as a single Zn treatment, no significant yield differences between the Zn treatments were found. However, for correcting Zn deficiency in spring cereals, application of Zn solution on the leaves two times, respectively 2 and 3 weeks after emergence, seems to give the best yield response.

Zinc concentrations in grain and young barley plants

As shown in Table 2, the Zn concentrations in barley grains were not influenced by the first year of Zn application. This is a common phenomena when a deficient crop responds to the treatment with high yield increases. For the following years the yield increases as influenced by Zn application were somewhat lower, and the Zn concentration in the grains showed an increasing tendency for increasing Zn applications.

In the year 1985 no Zn was applied to the barley crop, and no yield response was found this year for Zn treatments in previous years. The grain Zn concentration on the control plot had reached a level where Zn deficiency is no more expected.

The soil analysis (Table 5) shows that the soil Zn level on control plots had increased by nearly 45 per cent during the experimental period. A possible reason for this is that some Zn from Zn treated plots may have been transported to the control plots by soil tillage.

Assuming 20 ppm Zn in dry matter as a minimum range for normal development of barley, the 16 ppm Zn concentration found in young plants on the control plots in 1985 shows that the Zn supply from the soil may still be low.

In this experiment the Zn concentration found in wheat was somewhat higher than that for barley. This is in agreement with data published by BERGMANN (1983).

Potatoes

When a nutrient solution is sprayed on the foliage, some of the spray falls to the ground. The soil also receives a part of the nutrients applied through an increased nutrient concentration in plant residues. After some years a residual effect of the treatment is therefore expected.

In order to test a possible residual effect of Zn applied during the first three experimental years, no Zn was given to the potato crop in 1980.

As shown in Table 3, a significant increase in total tuber yield ($p < 0.001$), ware potato yield ($p < 0.01$), and dry matter yield ($p < 0.001$) were obtained. Zinc deficiency symptoms appeared on the leaves in some of the Zn-treated plots however.

The Zn concentration in the tubers was low, averaging 5.3 mg/kg dry matter in Zn-treated plots and 5.9 mg/kg in control plots. This finding shows that in spite of a significant residual effect on the yield, the Zn supply from the soil was insufficient.

In 1981 Zn solution was sprayed on the foliage according to the experimental plan, and a significant yield response was obtained. For treatments including the earliest spraying time, i.e. 4, 4+5, 4+6, and 4+5+6 weeks after emergence, the yields of ware potatoes were on average doubled compared with the yield on control plots (Table 3).

Zinc deficiency symptoms were visible four weeks after emergence, and the deficiency increased as time for Zn application was delayed. Applying Zn spray only once 5 or 6 weeks after emergence resulted in a substantial yield decrease. Zinc application at an earlier growth stage, before the deficiency symptoms were visible, could possibly have given a better yield response.

The Zn concentration in potato tubers increased due to the foliar application of Zn. The concentrations were 7.4, 11.3, 18.4 and 23.0 mg Zn/kg dry matter for the control, and for 1, 2 and 3 sprayings, respectively.

Red clover

In 1983, the seventh experimental year, the test crop was red clover. As no Zn was applied in that year, the differences in yield and plant Zn concentrations were due to a residual effect of Zn applied in previous years (Table 4).

Compared to the control a substantial yield increase was obtained at the first cut ($p < 0.01$), but, no significant yield differences were found for increased levels of Zn applied. At the second cut no significant yield response was found.

The Zn concentration in plants was lower in control plots, and lower at the second than at the first cut.

Zinc deficiency symptoms on the leaves, and a marked growth depression were observed on control plots early in the growing season. These observations were in agreement with the yields and with plant Zn concentrations at the first cut. Although Zn concentration in the regrowth was low on control plots, no deficiency symptoms and no yield depression were found at the second cut.

Changes in HCl-extractable soil Zn during the experimental period

Soil samples for analysis of HCl-extractable Zn were collected in autumn 1984, after eight experimental years. Application of Zn during the experimental period resulted in a marked increase in HCl-extractable soil Zn (Table 5). The increase in soil Zn in control plots, from 1.6 to 2.3 mg per kg airdry soil, is probably due to transport of soil from treated to untreated plots by soil tillage.

Conclusions

A zinc application to the plants by foliar spray was efficient in the experiments reported in correcting Zn deficiency on a Zn deficient heavily limed soil.

Foliar Zn treatments over some years resulted in an increase in plant available soil Zn. The results indicate that application of Zn to the soil could be a possible way of correcting Zn deficiency, even on heavily limed soils.

Summary

On a Zn deficient heavily limed sandy loam with 2.2 per cent organic Cm pH 7.5, and 1.6 ppm 0.2 HCl-exractable Zn, foliar application of 1 per cent solution of $ZnSO_4 \cdot H_2O$, 250 1/ha, increased spring cereal grain yields by 30 per cent, ware potato yield by 200 per cent, and potato dry matter yield by 30–40 per cent. The best treatment was spraying two times, for cereals 2 and 3 weeks afer emergence of seedlings, for potatoes 4 and 5 weeks after emergence of the plants, respectively.

After five years with foliar application of Zn solution, the 0.2 N HCl-extractable Zn content of the soil was increased by more than 100 per cent. The residual soil Zn caused a significant yield increase in red clover grown the first year without foliar application of Zn. The results indicate that residual Zn from foliar application accumulates in the soil, and is available to plants, even in heavily limed soils.

REFERENCES

AASEN, I. 1978   Mangan og sink i jord og planter (Manganese and zinc in soils and plants). Norsk Landbruk 1978: (5) 12–13, 38 (6) 14–15, 33, 38.

BERGMANN, W. 1983. Ernährungsstörungen bei Kulturpflanzen. Gustav Fischer Verlag. Stuttgart 1983. 614 p.

BRAEEKE, F.H. 1979. Boron deficiency in forest plantations on peatland in Norway. Meddr. Norsk inst. skogforsk. 35:213–236.

ELLIS, R. JR., J.F. DAVIS, and D.L. THURLOW 1964.
Zinc availability in calcareous Michigan soils as influenced by phosphorus level and temperature. Soil Sci. Soc. Amer. Proc. 28:83–86.

LINDSAY, WE.L. 1972. Inorganic phase equilibria of micronutrients in soils. In: Micronutrients in agriculture. Ed. committee: MORTVEDT, J.J., P.M. GIORDANO, and W.L. LINDSAY. Madison, Wisconsin USA 1972. pp 41–57.

NELSON, J.L., L.C. BOAWN, and F.G. VIETS, JR. 1959. A method for assessing zinc status of soils using acid–extractable zinc and "titratable alkalinity" values. Soil Sci. 88:275–283.

ODELIEN, M. 1967. Mikronaeringsstoffer, magnesium, svovel og kalsium. (Micronutrients, magnesium, sulphur and calcium). Ny Jord 54:49–65.

Table 1

Effects of Foliar Application of Zn on Grain Yield. Grain Dry Matter (tons/ha) on Control Plots and Relative Yield on Treated Plots when Control = 100.

Year	Species	Control	Spraying time, No. of weeks after emergence						
			1	2	3	1+2	1+3	2+3	1+2+3
1977	Barley	1.57	113	179	159	157	192	183	
1978	Oats	2.64	116	120	120	125	126	125	126
1979	Wheat	1.65	111	122	112	119	116	124	119
1982	Barley[1]	3.46	106	105	104	109	111	113	110
Mean		2.33	110	125	119	124	130	130	129

[1] Barley undersown red clover

Table 2

Zinc Concentrations (mg/kg of dry matter) in Grains and Young Barley Plants as Influenced by Foliar Applied Zn, Average for 1, 2 and 3 Sprayings

Year	Species	Control	Number of Zn sprayings		
			1	2	3
1977	Barley	8.7	9.3	8.9	
1978	Oats	10	13	13	20
1979	Wheat	22	22	23	25
1982	Barley	13	16	19	20
1985	Barley[1]	20	18	20	21
1985	Barley[2]	16	20	21	19

[1] No Zn was applied in 1985

[2] Plants sampled at the initiation of heading

Table 3

Effects of Foliar Application of Zn on Potato Yields
(tons/ha) on Control Plots, and Relative Yield on Zn
Treated Plots when Control = 100.

Year	Control	Spraying time, No. of weeks after emergence						
		4	5	6	4+5	4+6	5+6	4+5+6
		Tubers, total yield						
1980[1]	16.3	131	150	133	143	144	150	156
1981	26.1	140	129	126	144	135	133	142
		Ware potatoes (> 45mm)						
1980[1]	6.2	179	224	183	201	196	222	236
1981	11.3	201	156	131	207	189	183	213
		Dry matter						
1980[1]	4.1	135	155	137	149	149	154	161
1981	6.9	143	136	129	132	137	139	142

1) No Zn applied in 1980

Table 4

Effect of Zn Application in Previous Years on the Yield
of Red Clover and Plant Zn Concentration

No. of Zn applications per year	Dry matter yield tons/ha		Zn concentration in dry matter, mg/kg	
	1. cut	2. cut	1. cut	2. cut
0 (Control)	3.08	2.26	13.2	12.8
1	3.81	2.17	19.0	15.1
2	3.90	2.42	21.4	17.2
3	3.93	2.69	23.0	19.0

169

Table 5

HCl-extractable Soil Zn as Influenced by Zn Application on the Foliage during the Experimental Period

No. of Zn applications per year	Total amounts of Zn applied Kg/ha	HCl-extractable Zn mg/kg airdry soil
0 (Control)	0	2.3
1	4.87	3.0
2	9.70	3.2
3	11.39	3.8

APPLICATION DE ZINC AUX CEREALES, AUX POMMES DE
TERRE ET AUX TREFLES ROUGES POUSSANT SUR DES SOLS
FORTEMENT CHAULES DEPOURVUS DE ZINC

M. I. Aasen, Département de l'exploitation et
de la fertilité des sols de l'Université
agricole, Aas

RESUME

Les carences en oligo-éléments des cultures arables, des
légumes et des fruits sont bien connues en Norvège. C'est
essentiellement dans les sols tourbeux et dans les sols sablonneux
que l'on rencontre les carences en cuivre, auxquelles s'ajoutent sur
les sols argileux les carences en bore. La carence en manganèse est
toujours liée à un pH élévé et à un sol meuble. La carence en fer ne
se rencontre que sur les sols tourbeux acides. Depuis plus de
40 ans, on applique du bore, du cuivre et du manganèse pour corriger
ces carences. Il y a environ 35 ans que l'on a commencé à appliquer
en Norvège du molybdène en tant qu'élément nutritif.

La plupart du temps, la carence en zinc se rencontre sur des
sols calcaires ou fortement chaulés ayant un pH supérieur à 7. Avant
1970, un petit nombre seulement de carences en zinc avait été
signalé. Au cours des 15 dernières années, on a démontré qu'elle
existait dans les légumes, les fraises, les fruits, les céréales et
les pommes de terre. Dans des expériences en pots remplis de terre
dépourvue de zinc, on a trouvé aussi des carences dans les carottes,
le colza fourrager, les herbes et le trèfle rouge.

Sur des terres grasses sablonneuses fortement chaulées et
dépourvues de zinc, ayant un pH de 7,5 et contenant 2,2 % de carbone
organique, une application sur les feuilles des végétaux d'une
solution de Zn SO_4, H_2O, à concurrence de 250 litres par hectare
a fait augmenter le rendement des céréales de printemps de 30 % et
celui des pommes de terre de 39 %. La solution de zinc a été
appliquée deux fois, c'est-à-dire deux et trois semaines après
l'apparition des pousses de céréales et cinq et six semaines après
l'apparition des plants de pommes de terre.

Après cinq ans d'application d'une solution de zinc sur les feuilles des végétaux, la teneur du sol en zinc extractible -0,2 N HCl avait doublé entraînant un accroissement de 26 % du rendement en matières sèches du trèfle rouge. Les résultats de l'expérience montrent que le zinc qui n'est pas absorbé par les feuilles s'accumule dans le sol, et qu'il est utilisable par les plantes même dans les sols fortement chaulés.

MICRONUTRIENTS IN PLANT PRODUCTION IN FINLAND

Dr. J. Sippola, Institute of Soil Science,
Jokioinen, Finland

1.   Introduction

The mineral soils of Finland developed mainly from acid
magmatic rocks both during and after the glacial period.   Therefore
the total micronutrient content of the parent material of Finnish
soils is considered to be low.   Furthermore the weathering of soil
parent material in a temperate climate is slow.   Hence the
possibilities for micronutrient deficiencies were recognized long
ago.   The first field experiments on trace elements were established
in the late 1930s, and in the 1950s study of the total contents of
soil microelements on a larger scale was started.   Subsequently
there has been more emphasis on the determination of plant available
contents.   The utilization of micronutrients in fertilizers is based
on soil testing, but to satisfy the health needs of cattle in
certain areas fertilizers supplemented with micronutrients are used.

2.   Total micronutrient content of soils

An intensive spectrographic survey of total micronutrient
contents was started in the 1950s.   The results showed that the
average total micronutrient content of Finnish soils does not
greatly differ from that of the earth's crust (table 1).   However,
great differences in the trace element content between soil types
were found.   Texture was the decisive factor in determining the
content of trace elements.   The trace element contents of heavy clay
are multiples of the corresponding amounts in sand soils.   For
molybdenum and cobalt the diference is from 4 to 5 fold.   This is
due to micaceous clay minerals in fine textured soils versus high
amounts of quartz in sand soils.   Moreover the micronutrient content
of some peat soils is low, especially in the case of Spaghnum peats,
whose use in agriculture, however, is very limited.   Total content
of soil micronutrients is only a general index for plants whereas
soluble amounts are more important.   Recently therefore more
emphasis has been on the determination of extractable contents.

3.   Methods of soil micronutrient extraction

In Finland, there has been a tendency to use one universal extractant for all micronutrients. Only in the case of boron has a specific extractant, the well known hot-water extraction method, been accepted for general use. The first attempts to test a single extractant for several available micronutrients were made using acid ammonium acetate, pH 4,65. The same extractant is used for macronutrient extraction in soil testing in Finland. However, this extractant is very weak and the amounts of micronutrients extracted are in some cases so low that their determination is difficult. One has to rely on pre-concentration before analysis which is time-consuming and costly.

The results for the determinations of available micronutrients using acid ammonium acetate varied between soil types much less than those for the total content (Table 2). Contents extracted were a few per cent of the total content, or less. Especially the extracted amount of copper was very low. Therefore to avoid difficulties in the determinations the efficiency of this same extractant was increased by adding EDTA to a concentration of 0.02 M. The results of a comparative investigation on these methods showed that the EDTA addition greatly increased the extraction of most micronutrients (Table 3). The effect of EDTA addition was, however, quite different for various micronutrients. For example zinc was slightly affected because of the low stability of the Zn-EDTA complex at the pH 4.65. Contents of manganese and cobalt were increased by a factor of two. The contents of molybdenum were about five times higher and those of copper more than twenty.

The AAAc+EDTA extraction method has been accepted for use as the reference method of the FAO European Co-operative Network on Trace Elements. In Finland, this method is now used in soil testing for micronutrients. Based on the results of a pot experiment, correlation of the plant uptake of various micronutrients with those extracted from soil with AAAc+EDTA was closer than that from the contents of the 2 N HCl extraction formerly used to determine Cu and Zn by the Soil Testing Survey (Table 4). Correlation of the results of the well known DTPA method were slightly better but the difference was not significant. In the case of manganese the use of the correction factor based on soil pH improved the correlation coefficient to .47 when AAAc+EDTA was used and to .43 by DTPA.

Critical limits for interpretation of the AAAc+EDTA results were based on results obtained from pot experiments, on the distribution of results in surveys made, and on comparing these results with earlier classifications (Table 5).

4. Contents of extractable micronutrients in soils

Boron. Deficiency of boron is very common and this is also indicated by the amount of soils whose boron contents are below the critical limit 0.2 mg/1 soil (Fig. 1). Their total amounts to about one fourth of the sampled fields. No soil type was especially low in boron and deficiency was met throughout the country. Therefore, the decision to add boron to all chemical fertilizers has been a sound one.

Copper. About one fifth of the samples were deficient in copper according to the critical limit of 1 mg/1 soil. Contents were low especially in Carex peat and sandy soils. The copper content in one quarter of the samples in these groups was below the critical limit. The greatest copper content was found in clay soils. By region, low contents were common in Northern Finland where peat soils are common and in Central Finland where coarse textured mineral soils dominate.

Manganese. About 5 per cent of the samples contained manganese below the critical limit of 6 mg/1. Therefore manganese deficiency does not seem very likely. Because of the dependence of manganese availability on soil pH heavy liming, for example as in sugarbeet cultivation, may cause deficiency. Furthermore in some acid soils total reserves may be low and in intensive cultivation employing the heavy use of nitrogen fertilizers, a genuine manganese deficiency may develop. Such indications have been found in grassland cultivation in Northern Finland.

Molybdenum. Slight problems with respect to molybdenum may be expected because only a very few samples contained less than 0.01 mg/1 molybdenum. Problems with molybdenum are associated with the change in soil pH when acidifying fertilizers are used.

Zinc. About 10 per cent of the samples contained zinc below the critical limit of 1 mg/1, and zinc deficiency has not been clearly identified. However, in the interest of animal health a fertilizer containing zinc has been introduced.

5. The use of micronutrient fertilizers

Fertilizers containing boron, copper and manganese are the micronutrient fertilizers most used on cultivated fields. Nearly 800 tons of boron are applied yearly; an average of 350 g/ha (Table 6). Special crops such as sugarbeet are fertilized more heavily, but for cereals this rate of application is very high. Consequently, the soil boron content has increased accordingly so that at the beginning of the current fertilizing year the boron content of most common fertilizers has been lowered by 40 per cent from 0.05 per cent to 0.3 per cent, respectively.

In addition, the application rate of copper seems to be high
when compared to crop removal. Copper is mostly applied on the
copper deficient soils of Northern Finland but a large quantity of
copper is also added to special fertilizers for horticultural use
and has been included in the above figure. Farmyard manure may also
have a high copper content due to the mineral supplementation of
feeds. Some copper is deposited via rainfall, thus the copper
balance of fields is positive.

Almost 200 g/ha of manganese is supplied in mineral
fertilizers. Via cattle manure more than this amount is brought
onto the fields. Small amounts of manganese are deposited onto
fields by atmospheric deposition. However, total reserves of soil
manganese are good and deficiency should not become too common
unless pH is radically raised by liming.

No statistics are yet available on the use of zinc in
fertilizers, as beside the Zn-containing fertilizers for
horticultural use, the first Zn-fertilizer for increasing the Zn
content of cattle feed has been on the market for only one year.
The average balance, however, appears to be positive without any
additional amount as the Zn in manure replaces its removal by a
cereal crop. Also rain deposition is an important source of soil
zinc in Finland.

6. Selenium fertilization

The problems caused by selenium deficiency in animal husbandry
and discussions on the possible detrimental effects of a low
selenium level in the Finnish diet led to the supplementation of
selenium of chemical fertilizers in 1984. The rates added are
16 mg/kg in those fertilizers used for cereals and 6 mg/kg in
fertilizers used for grassland and silage. The first results show
that due to this addition the selenium intake of the Finnish
population has doubled (Yläranta 1986).

176

References

Kemira Oy 1985. Lannoitteiden myynnin jakautuminen
matalouskeskusalueittain lannoitusvuonna 1984–85. 16 p.

Kemppainen, E. 1984. Karjanlannan ravinnepitoisuus ja syyt sen
vaihteluun. SITRA, julk. 11.

Sillanpää, M. and Lakanen, E. 1966. Readily soluble trace elements
in Finnish soils. Ann. Agric. Fenn. 5: 298–304.

Sillanpää, M., Lakanen, E., Tares, T. and Virri, K. 1975.
Hivenaineiden uutto EDTA:lla tehostetulla happamalla
ammonium-asetaatilla suomalaisista maista. Kehittyvä Maatalous
21: 3–13.

Sippola, J. and Erviö, R. 1986. Comparison of some soil testing
methods for micronutrients. Transact. ISSS 13th Congr. Vol. III:
972–973.

Sippola, J. and Tares, T. 1978. The soluble content of mineral
elements in cultivated Finnish soils. Acta Agr. Scand., Suppl. 20:
11–25

Soveri, J. 1976. Epäpuhtauslaskelmista Suomessa 1975–1976,
sadevesi- ja lumianalyysien avulla arvioituna. Ympäristö ja terveys
7: 837–847.

Viljavuuspalvelu Oy. 1986. Viljavuustutkimuksen tulkinta
peltoviljelyssä. 63p.

Vuorinen, J. 1958. On the amount of minor elements in Finnish
soils. J. Scient. Agric. Soc. Finl. 30: 30–35.

Yläranta, T. 1986. Seleenilannoituksen vaikutus ihmisten ja
eläinten seleenin saantiin. Lecture 11.2.1986.

Table 1

Total Content of Micronutrients by Soil Types, mg/kg
(VUORINEN 1958)

Element	Heavy clay	Silt	Finer fine-sand	Fine-sand	Mould	Carex peat
Cu	36	23	17	15	26	16
Mn	780	1 000	650	420	480	350
Mo	12	12	4	7	8	6
Zn	70	60	30	20	30	20

Table 2.

Average Micronutrient Contents Extracted with Acid Ammonium
Acetate, pH 4.65, from Various Soil Types, mg/1
(SILLANPAA and LAKANEN 1966).

Element	Clay	Silt	Fine-Sand	Mould	Carex peat
Cu	0.27	0.16	0.18	0.26	0.14
Mn	22	26	22	34	27
Mo	0.005	0.005	0.006	0.017	0.020
Zn	2.6	2.8	2.9	3.3	3.0

Table 3.

Extractability of Micronutrients with Acid Ammonium Acetate,
pH 4.65 (AAAc), and with 0.02 M EDTA, mg/1 soil
(SILLANPAA et al. 1975).

Extractant	Zn	Mn	Co	Mo	Cu
AAAc	3.03	25	0.14	0.009	0.20
AAAc+EDTA	2.83	39	0.30	0.052	4.68

Table 4.

Correlation between the Squareroot of Soil Micronutrient Content Extracted with Different Methods and their Uptake by Ryegrass
(SIPPOLA and ERVIO 1986).

Element	AAAc+EDTA	DTPA	Ashing+ 2 N HCl	MgSO4
Cu	.65	.69	.56	
Mn	.23	.28		.54
Mn+pHcorr	.47	.43		
Zn	.86	.87	.16	

Table 5.

Interpretation of Soil Test Results for Micronutrients
(VILJAVUUSPALVELU OY 1986).

The range limit values shown belong to the upper class

Element	Poor	Rather poor	Fair	Satis-factory	Good	Very good	Possible excessive
Boron, mg/l							
Clay soils	.3	.5	.8	1.2	1.7	2.5	
Other soils	.2	.4	.6	.9	1.3	2.0	
Copper, mg/l	1.0	1.5	2.0	5.0	10	20	
Manganese pH-corr	6	12	25	75	250	1000	
Zinc, mg/l	1.0	1.5	2.0	6.0	20	50	
Molybdenum mg/l	.01	.02	.03	.06	.2	.5	

Table 6.

The Use of Micronutrients in Mineral Fertilizers in 1984-86, in Manure, Atmospheric Deposition and Removal by Barley, g/ha
(KEMIRA OY 1985, KEMPPAINEN 1984, SOVERI 1976)

Item	B	Cu	Mn	Zn
Fertilizers	349	314	189	
Manure		30	250	200
Rain		1	2	4
Uptake by barley	12	22	500	140

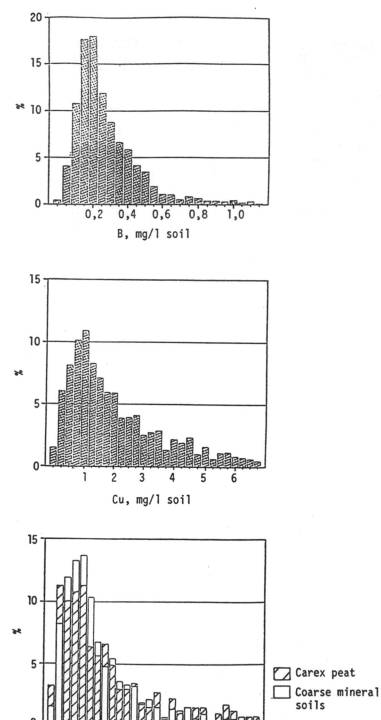

Fig. 1a. Distributions of soil extractable boron and copper.

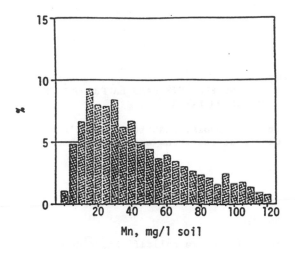

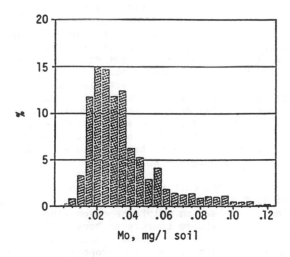

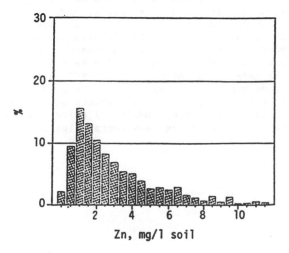

Fig. 1b. Distributions of soil extractable manganese,
molybdenum and zinc.

LES OLIGO-ELEMENTS DANS LA PRODUCTION
VEGETALE EN FINLANDE

Dr. J. Sippola, Institut de pédologie,
Centre de recherche agricole, Jokioinen

RESUME

C'est principalement la solidification de roches magmatiques
acides pendant la période glaciaire qui a donné naissance aux sols
minéraux de la Finlande, d'où une très faible teneur globale en
oligo-éléments ("micronutriments") des sols finlandais. Le risque
de carences en oligo-éléments étant reconnu depuis longtemps les
premières tentatives pour étudier l'effet des oligo-éléments
remontent à la fin des années 30. C'est dans le courant des
années 50 que l'étude de la teneur globale des sols en
oligo-éléments a été lancée à grande échelle. Les résultats obtenus
ont montré que les variations des teneurs globales en oligo-éléments
entre les différents types de sols étaient surtout dues à des
différences de texture. Depuis, on s'est davantage attaché à
déterminer la teneur des plantes en oligo-éléments, exercice pour
lequel on a mis au point un extracteur (acétate d'ammonium + acide
éthylène diamine tétracétique - EDTA).

Une étude des terres cultivées dans toute la Finlande a révélé
qu'entre 10 et 20 % des sols manquaient de bore ou de cuivre.
Les carences en manganèse, en molybdène ou en zinc se sont révélées
moins fréquentes. De faibles teneurs en bore sont courantes dans
tout le pays, aussi les engrais azotés, potassiques et phosphatés
(NKP) contiennent-ils du bore dans des proportions suffisantes pour
remplacer les déperditions annuelles. Les carences en cuivre sont
surtout relevées dans les sols à texture grossière et dans les
tourbes. Une carence en manganèse se rencontre occasionnellement
dans les sols fortement chaulés. Les faibles teneurs en zinc sont
rares, mais pour augmenter celles des aliments pour animaux un
engrais contenant du zinc a été mis sur le marché. Selon la culture
et les sols, la teneur moyenne des engrais NKP en oligo-éléments est
de 350 grammes de bore, 300 grammes de cuivre et 200 grammes de
manganèse à l'hectare. La découverte récente de la pauvreté en
sélénium des sols finlandais a conduit à l'adjonction de Se (dans
des proportions de 6 à 16 mg/kg) aux engrais les plus couramment
utilisés.

L'agriculture moderne a généralement pour conséquence
l'appauvrissement des sols en oligo-éléments, mais cet effet est
compensé par l'introduction d'oligo-éléments dans les engrais et par
les apports du fumier des bovins qui ont absorbé des oligo-éléments
avec les sels minéraux qui leur sont administrés dans les élevages
intensifs. Les tests destinés à contrôler la teneur des champs en
oligo-éléments jouent un rôle de plus en plus important.

THE MINERALOGICAL COMPOSITION AND SOME
PHYSICAL AND CHEMICAL PROPERTIES OF CLAY
AND SILT SAMPLES RELATED TO THEIR SPECIFIC
COPPER ADSORPTION

Mr. J. Dissing Nielsen
State Laboratory for Soil and Crop Research,
Lyngby, Denmark

INTRODUCTION

The continuing contamination of soils with heavy metals through
the application of fertilizers, sludge and by air pollution and the
recognition of the associated danger to the environment give
justification for an increasing number of investigations into the
reactions of heavy metals such Zn, Cu, Cd, Pb and Ni in soils and
with various soil fractions (Tjell, 1978; Assaad & Nielsen, 1984,
1985; Christensen, 1984; Tiller et al., 1984; Debosz et al.,
1985; Korcak & Farming, 1985; Rothbaum et al., 1986). The
particle size fractions of clay, silt and humus are particularly
important in mineral soils for the adsorption of heavy metals. They
can be useful as micronutrients as well as damaging in relatively
high concentrations which appear as artificial compounds not
naturally present in the soil.

The clay particle fraction contains crystalline soil minerals,
primarily clay minerals. Table 1 shows the composition of clay and
silt from a typical Danish surface soil sample. The prevailing clay
mineral is illite, but vermiculite, smectite, chlorite and kaolinite
are also usually found in Danish soils. Quartz and feldspars are
equally present in the clay fraction but these minerals do not
possess the structure of clay minerals and are only weak adsorbers
of Cu and other cations. The metaloxidhydroxides of Al, Si and Fe
are non-crystalline and important adsorbers of most heavy metals.

The silt particle fraction only contains few clay minerals and
is composed mainly of quartz and feldspars. The humus fraction is a
mixture of high-molecular C compounds linked to the soil minerals
and with a beneficial effect on soil structure. Humus is
particularly linked to clay – 50 to 70 per cent – and silt –
20 to 35 per cent – (Christensen, 1985). A large part of CEC of the
soil is due to humus compounds and particularly for sandy soils, the
humus content is important for soil fertility. Humus is forming a
complex with most heavy metals and reduces the plant availability

and permeability of heavy metals to drain and ground water. In Denmark, deficiency of Cu and Zn is mainly observed in blank sandy moorland soils in the western part of Jutland.

The intention of this study was to obtain further knowledge on Cu adsorption in soils of various textures and mineralogical composition known from other investigations (Møberg & Nielsen, 1986). Samples with and without humus were investigated.

Sampling and methods

The samples were from the A-horizon of soil profiles. Table 2 shows the classification and texture of the soils investigated. The samples were air-dried before laboratory treatment. A soil quantity corresponding to 10 g of clay was weighed and fractionated into clay, silt and sand. Humus was removed by oxidation with hydrogen-peroxide. The samples were dispersed in 0.002 M $Na_4P_2O_7$ and the sand particle fraction was separated from clay and silt after sedimentation. Clay and silt were separated in a particle size centrifuge (Slanter & Cohen, 1962). After separation of clay and silt, the clay fraction was saturated with NH_4Cl or $Ca(NO_3)_2$ and precipitated. Silt was also saturated with NH_4 or Ca-ions and both particle size fractions were washed with ethanol until being free of Cl or NO_3-ions. The precipitated samples were air-dried.

The specific adsorption of Cu was studied in the NH_4 or Ca-saturated clay and silt samples. To measure the adsorption of Cu, a series of 0.1 g samples were placed in centrifuge polyethylene tubes and 5 ml of a graded concentration of Cu (from 0 to 100 ppm) as $CuCl_2$ in 0.05 M NH_4Cl or 0.01 M $Ca(NO_3)_2$ were added. The samples were shaken on a reciprocation shaker three times during 48 hours of equilibration at room temperature (circa 20°C). The tubes were centrifugated at 8000 rpm for 5 minutes, and the Cu content of the supernatant was measured by atomic spectrophotometry. The adsorbed Cu was determined by a washing procedure using 0.01 M $Ca(NO_3)_2$ (Tiller et al., 1984). The amount of Cu remaining in the clay and silt fraction was defined as specifically adsorbed Cu.

The pH of the solutions was estimated after equilibration with Cu. The variance ranged from 5.5 to 6.5 with the lowest values for the highest Cu concentrations. In this pH range the influence of pH on Cu adsorption was found to be small (Assaad & Nielsen, 1984). The influence of humus on Cu adsorption was studied in samples not oxidized with hydrogenperoxid.

Results and discussion

The cation exchange capacity (CEC) was estimated at pH 7.0 in clay minerals, clay and silt fractions. CEC increased in clay minerals in the order kaolinite - illite - smectite - vermiculite. It was shown by Hilling et al., (1964) and confirmed in the present experiment that CEC for clay-humus is lower than the added CEC's for separate clay and humus fractions. The theory was submitted by Hilling et al., (1984) that a separation of clay and humus is opening sites for exchange. These authors also calculated the CEC contribution of clay and humus at different pH values from the relationships between CEC and the clay and humus content. CEC of humus increased strongly with a pH increase from 2.5 to 8.0 i.e. by a factor of 6 whereas for clay in the same pH range CEC increased by a factor of 1.7.

The results in Table 3 show that the influence of humus on CEC is most pronounced in sandy soils in accordance with a higher C content in the clay and silt fractions of this soil type. For clay soils CEC in silt and silt-humus is 2 meq per 100 g sample. Sandy soils are richer in C in clay and silt particle fractions than clay soils. The silt particle fraction of sandy soils showed some CEC disappearing after oxidizing the sample.

The surface area of clay and silt samples without humus was estimated after saturating the samples with H_2O and drying with P_2O_5. The surface area was greater for clay compared to silt fractions. Only small differences were found between various localities and soil types.

Results regarding the specific adsorption of Cu in some clay minerals are shown in Figure 1. Kaolinite with a 1:1 structure shows less Cu adsorption than illite, smectite and vermiculite with a 2:1 structure. The structure of kaolinite causes an almost neutral molecule with only few sites for adsorption of cations. The clay minerals with a 2:1 structure possess a more negative charge and more sites for Cu adsorption than kaolinite. The structure of illite is particularly adapted for K-ions. Illite extends only a small amount with water (non-quelling) contrary to vermiculite and smectite. The specific adsorption of Cu is less for illite than for the quelling clay minerals.

Divalent cations e.g. Ca extend the clay structure more than monovalent cations e.g. NH_4. Figure 2 shows the specific Cu adsorption for smectite saturated with NH_4 or Ca. The adsorption is greater for the Ca saturated than for the NH_4 saturated sample. For kaolinite without extending structure the Ca adsorption was not changed with different ion coverage.

For some of the clay and silt fractions the Cu adsorption was compared in NH_4 - and Ca-saturated samples. More Cu was adsorbed specifically in Ca than in NH_4 saturated soil fractions, but the ratios were not changed by different ion saturation. Calculations of the correlation between specific Cu adsorption and soil mineralogy were only made for the NH_4 saturated samples.

Cu adsorption in clay minerals and in clay and silt particle size fractions showed the same shape of the curve as found for whole soils (Assaad & Nielsen, 1984). The adsorption of Cu conformed to the linear form of the Langmuir equation:

$$C/\underline{x} = \frac{1}{a.b} + \frac{C}{a}$$
$$\quad m$$

$\underline{x}$ = The amount of Cu adsorbed by unit weight
$\quad m$

C = Equilibrium Cu concentration in solution

$\underline{a}$ = The Langmuir adsorption maximum

$\underline{b}$ = The Langmuir "bonding term" related to bonding energy

Table 3 shows Cu max. adsorption $\underline{a}$ and bonding term $\underline{b}$ for clay minerals and for clay and silt particle size fractions with and without humus.

For pure clay minerals $\underline{a}$ was lower than for the clay fractions separated from soil samples in spite of the latter's containing some inactive material such as quartz and feldspars. This unexpected result could be due to a certain content of silty material - 10 to 20 per cent - in the clay minerals and to high affinities for Cu of the oxidhydroxides contained in the separated clay fractions. The Cu $\underline{a}$ value increased in the same order as for CEC; kaolinite - illite - smectite - vermiculite.

A comparison of the Cu $\underline{a}$ values of clay and silt shows that they are highest for clay which seems obvious because of the clay minerals in the clay fraction and its higher surface area.

Removal of humus from clay and silt reduces the Cu $\underline{a}$ value for sandy soils and for silt from clay soils. On the other hand, the last soil type showed the highest Cu $\underline{a}$ value after removal of humus from the clay. The increased Cu adsorption in clay fractions free of humus for clay soils could be caused by treatment effects (Møberg & Nielsen, 1986).

Figure 3 shows that an increasing content of C in clay and silt results in an increasing adsorption of Cu. This is in accordance with practical results according to which organic soils are strong adsorbers of Cu and Cu deficiency is particularly related to organic sandy soils.

The surface area is probably important to the Cu a value but owing to the limited estimations of surface area the correlation was not calculated. It can be mentioned though that the highest surface area was found in the sample with the highest Cu a value.

The expression for bonding energy b varies with the soil type. There was a tendency towards higher Cu b values for samples containing humus and for clay compared to silt fractions.

Table 4 shows the mineralogical composition of clay samples from the soils used in the experiment. The same minerals are generally present in all samples and illite is the most prevailing mineral. The content of vermiculite, smectite, chlorite and kaolinite differs between locations. Insignificant correlations were found between soil texture and the content of these minerals. However, the clay samples from sandy soils contained more quartz, feldspars and metaloxidhydroxides than clay soils.

Table 4 also shows the correlations between Cu a values and the mineralogical composition of clay samples. A significantly negative correlation was found between Cu a values and the content of illite. The investigation with pure clay minerals showed that Cu adsorption was less for illite than for the quelling clay minerals smectite and vermiculite. Negative correlations were also found between the content of quartz plus feldspars and Cu a values.

A positive correlation was found between the content of metaloxidhydroxides and Cu a values in spite of CEC being lower for these compounds than for smectite and vermiculite. In particular the most weathered sandy soils showed a high binding of Cu. It may be expected that these soils will also have a high capacity for binding other heavy metals.

On account of the relatively high Cu binding in smectite and vermiculite a positive correlation between Cu a values and the content of these clay minerals could be expected but the expected relationship was not confirmed. More experiments will be needed to explain this deviation. As could be expected correlations were not found between the content of chlorite or kaolinite and the Cu a values.

Conclusions

1.   Cu binding in soil fractions is related to the inorganic part of clay and silt. In silt Cu binding takes place primarily in humus.

2.   Clay fractions are the principal adsorbers of Cu compared to the other soil texture fractions.

3.   The results showed only weak relationship between Cu adsorption and the mineralogical composition of clay samples. However, for the dominating mineral illite and also for quartz plus feldspars a negative correlation was found.

4.   The content of Al-Fe-oxidhydroxides was positively correlated with Cu adsorption.

5.   The results of the experiment indicated that greater knowledge of the mineralogical composition of soil samples will improve the possibility of predicting their Cu adsorption.

Table 1

Estimated Composition of the Clay and Silt Fraction from Roskilde. Horizon Ap (0-25 cm)

Component	Clay	Silt
	Percentages	
Na + Ca feldspar	2-5	15-19
K-feldspar	1-4	12-16
Illite	40-44	
Vermiculite	13-17	
Chlorite	4-7	
Smectite	4-7	
Kaolinite	6-9	2-5
Quartz	10-14	59-63
Mica		3-6
Amphiboles		0-3

Table 2

Classification and Texture of Soils, Particle Size in mu

Locality	Classification	Percentage				
		2	2–20	20–200	200	Humus
Møjer	Mollic Fluvaquent, fine to coarse loamy, mixed, mesic	14.8	15.5	66.8	1.6	1.3
Årslev	Typic Agrudalf, fine loamy, mixed, mesic	12.0	16.7	42.1	26.7	2.5
Flakkebjerg	Typic Agrudalf, fine loamy, mixed, mesic	14.7	24.7	35.3	22.6	2.5
Roskilde	Typic Agrudalf, fine loamy, mixed, mesic	11.5	17.5	41.8	26.8	2.4
Rønhave	Typic Agrudalf, coarse loamy, mixed, mesic	12.0	21.6	40.2	23.0	3.2
Foulum	Typic Hapludult, coarse loamy, mixed mesic	6.9	12.8	49.2	27.5	3.6
Borris	Typic Haplothord, sandy, mixed mesic	5.6	8.8	45.4	38.0	2.2
Tylstrup	Typic Udipsamment, mixed, mesic	2.8	7.0	10.8	76.5	2.9

Table 3

Cation Exchange Capacity (CEC), C-content, Surface Area and Langmuir
Constants for Specific Copper Adsorption of Clay and Silt Samples

Locality and designation	Treatment	CEC (meq./100 g soil)	C-content (%)	Surface area (m²/g)	a (ug/g)	b (g/ml)
Borris	Clay+humus	43	10.08		58	2.2
Sandy	Clay	43			53	2.4
	Silt+humus	8	5.76		37	2.5
	Silt	2			15	2.0
Tylstrup	Clay+humus	74	11.70		76	0.5
Sandy	Clay	52				
	Silt+humus	27	9.88		57	0.5
	Silt	2				
Kaolinite		7			8	2.4
Illite		30			27	0.4
Smectite		100			15	2.4
Vermiculite		150			41	0.3

Table 3 (continued)

Locality and designation	Treatment	CEC (meq./100 g soil)	C-content (%)	Surface area (m²/g)	$\underline{a}$ (ug/g)	$\underline{b}$ (g/ml)
Højer	Clay+humus	46	5.16		47	0.8
Clay	Clay	41		63	41	1.2
	Silt+humus	2	4.30		30	0.5
	Silt	2		14	9	0.3
Arslev	Clay+humus	36	6.72		54	2.1
Sandy clay	Clay	50		67	72	0.7
	Silt+humus	2	3.54		21	0.2
	Silt	2		10	12	0.8
Flakkebjerg	Clay+humus	42	5.22		43	2.4
Sandy clay	Clay	50		55	63	0.7
	Silt+humus	2	2.61		23	0.2
	Silt	2		9	12	0.8
Roskilde	Clay+humus	51	6.48		44	2.2
Sandy clay	Clay	51			54	2.0
	Silt+humus	2	6.48		20	0.3
	Silt	2			8	0.6

Table 3 (continued)

Locality and designation	Treatment	CEC (meq./100 g soil)	C-content (%)	Surface area (m²/g)	a (ug/g)	b (g/ml)
Rønhave	Clay+humus	47	6.44		43	2.4
Sandy clay	Clay	51		62	52	0.5
	Silt+humus	2	3.62		21	0.2
	silt	2		8	5	0.3
Foulum	Clay+humus	70	11.40		78	4.4
Sandy	Clay	40			54	1.7
	Silt+humus	22	7.40		39	5.1
	silt	1			13	0.8

Table 4

Mineralogical Composition of Clay Samples (per cent) and Correlation r to Cu-adsorption Maximum a

Locality	Illite	Vermiculite	Smectite	Chlorite	Kaolinite	Quartz + feldspars	Metaloxid hydroxides
Højer	25	8	20	10	14	19	7
Arslev	25	12	11	8	14	22	6
Flakkebjerg	33	13	1	7	15	21	5
Roskilde	42	15	2	5	7	21	5
Rønhave	33	5	21	2	21	12	7
Foulum	26	19	11	5	2	36	7
Borris	23	15	10	0	5	27	14
Tylstrup	18	18	1	4	13	33	13
"r"	-0.35*	0.25	-0.14	0.13	0.11	-0.25*	0.38*

* Significant at the 5 per cent level

Literature

Assaad, F.F. and Nielsen, J.D. (1984): A thermodynamic approach for copper adsorption on some Danish arable soils. Acta Agric. Scand. 34: 377–385.

Assaad, F.F. and Nielsen, J.D. (1985): Adsorption of zinc in selected soils from Denmark. Acta Agric. Scand. 35: 48–54.

Christensen, B.T. (1985): Carbon and nitrogen in particle size fractions isolated from Danish arable soils by ultrasonic dispersion and gravity sedimentation. Acta Agric. Scand. 35: 175–187.

Christensen, Th. H. (1984): Cadmium soil sorption at low concentrations: I. Effect of time, cadmium load, pH and calcium; II. Reversibility effect of changes in solute composition and effect of soil aging. Water air and Soil Pollution 21: 105–125.

Debosz, K., Babich, H. and Stotzky, G. (1985): Toxicity of lead to soil respiration: Mediation by clay minerals, humic acids and compost. Environmental Contamination and Toxicology 35: 517–524.

Korcak, R.F. and Farming, D.S. (1985): Availability of applied heavy metals as a function of type of soil material and metal source. Soil Science 140: 23–33.

Helling, C.S., Chesters, G. and Corey, R.B. (1964): Contribution of organic matter and clay to soil–cation exchange capacity as affected by the pH of the saturating solution. Soil Science Soc. of Am. Proc. 28: 517–520.

Møberg, J.P. and Nielsen, J.D. (1986): The constituent composition of soils from Danish State Agricultural Experimental Stations. Danish Journal of Plant and Soil Science (in print).

Rothbaum, H.P., Goguel, R.L., Johnstone, A.E. and Mattingly, G.E.G. (1986): Cadmium accumulation in soils from long–continued applications of superphosphate. Journal of Soil Science 37: 99–107.

Slanter, C. and Cohen, L. (1962): A centrifugal particle size analyzer. Journal Science Instrum. 39: 614–617.

Tiller, K.G., Gerth, J. and Brummer, G. (1984): The sorption of Cd,
    Zn and Ni by soil clay fractions: Procedures for partitioning
    of bound forms and interpretation. Geoderma 34: 1-16.

Tjell, J.C. and Hovmand, M.F. (1978): Metal concentrations in Danish
    arable soils. Acta Agric. Scand. 28: 81-88.

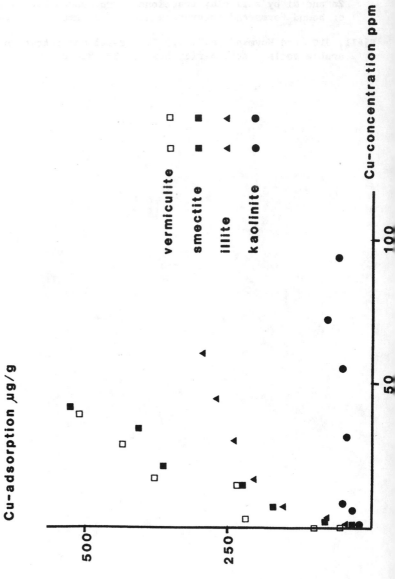

Figure 1

Specific adsorption of Cu in clay minerals

Specific adsorption of Cu in smectite saturated with NH_4 or Ca-ions

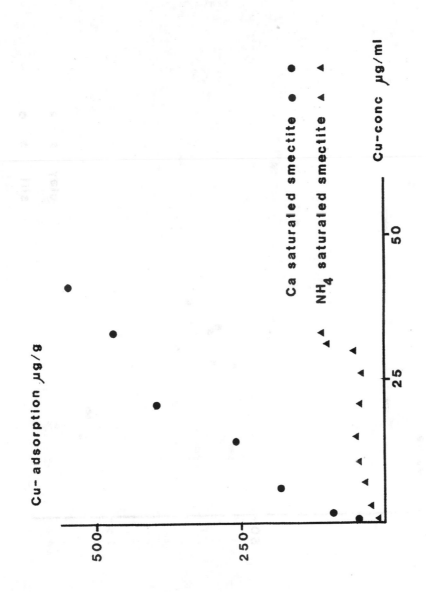

Cu-adsorption $\mu g/g$

Ca saturated smectite ●

NH_4 saturated smectite ▲

Cu-conc $\mu g/ml$

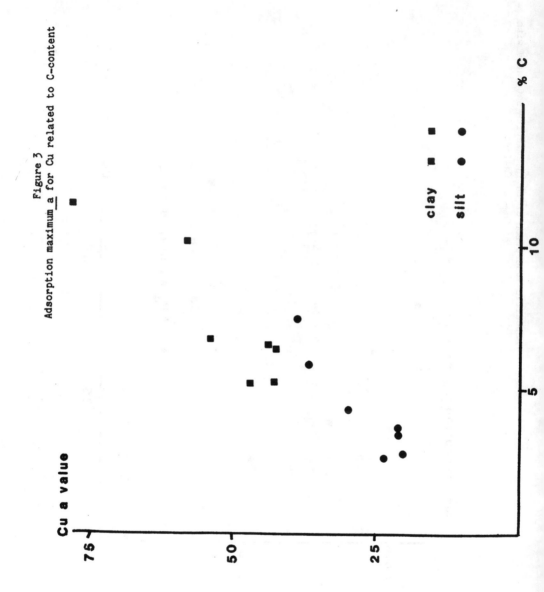

Figure 3
Adsorption maximum a for Cu related to C-content

LA COMPOSITION MINERALOGIQUE D'ECHANTILLONS
D'ARGILE ET DE LIMON ET QUELQUES PROPRIETES
PHYSIQUES ET CHIMIQUES DE L'ABSORPTION
SPECIFIQUE DU CUIVRE PAR CES SUBSTANCES

M. J. Dissing Nielsen, Laboratoire national
de recherche sur les sols et les cultures,
Lyngby

RESUME

Les métaux lourds sont absorbés par certaines fractions du sol,
essentiellement l'argile, le limon et l'humus. Dans l'expérience
décrite ici, on a étudié l'absorption spécifique du cuivre par les
minéraux argileux et par des fractions d'argile et de limon de
dimension particulaire dans certains sols. On a calculé le rapport
entre l'absorption du cuivre et la composition minéralogique des
fractions d'argile de dimension particulaire. Ces calculs n'ont
permis de trouver que des liens ténus entre elles. Le rapport a été
négatif pour la teneur en illite et en quartz feldspathique, mais
positif pour les hydroxydes métalliques. La partie de l'argile et
en particulier le limon contenant de l'humus joue un rôle
prépondérant dans l'absorption spécifique du cuivre.

TENEURS EN FER, MANGANESE, CUIVRE, ZINC,
BORE ET LITHIUM DANS LES GRAINS DU BLE
D'AUTOMNE ET DANS LES TUBERCULES DE
POMMES DE TERRE (RESULTATS PRELIMINAIRES)

MM. J.P. Quinche, A. Maillard et W. Reust,
Station fédérale de recherches agronomiques
de Changins, Nyon, Suisse

INTRODUCTION

Un nouveau projet de recherches intitulé "Composition minérale
des productions végétales" a été mis en route récemment à la Station
fédérale de recherches agronomiques de Changins. Dans ce rapport,
nous présentons les méthodes d'analyses utilisées et les premiers
résultats obtenus chez le blé et la pommes de terre lors des
récoltes de 1985. Le but de ces travaux est d'étudier l'évolution
de la composition minérale des aliments d'origine végétale en
fonction du degré d'intensification des cultures.

CONDITIONS CULTURALES ET PREPARATION DES ECHANTILLONS

Les caractéristiques des lieux d'essai et les conditions
culturales figurent au tableau 1. Chacune des variantes
expérimentales était répétée 4 fois dans le champ d'essai. Lors des
récoltes, des échantillons de grains de blé et de tubercules de
pommes de terre ont été prélevés dans chaque répétition, puis
mélangés pour fournir un échantillon moyen par variante, en vue des
analyses chimiques.

Les grains de blé ont été séchés dans une étuve à 105°C
durant une nuit, broyés grossièrement dans un mortier de porcelaine,
puis moulus finement avec un moulin à billes d'agate pendant 15
minutes (Pulverisette Fritsch).

Les tubercules de pomme de terre ont été lavés à l'eau
déminéralisée, essuyés puis découpés en menus fragments à l'aide
d'un couteau en acier inoxydable; des prises d'environ 200 g de ce
matériel ont été introduites dans des ballons de verre, congelées
dans de l'azote liquide puis lyophilisées. Après pesée pour la
détermination de la matière sèche, les pommes de terre lyophilisées
ont été pulvérisées à l'aide d'un moulin à hélice (Moulinette
Moulinex)

METHODE DE DOSAGE DES ELEMENTS Fe, Mn, Zn, Cu, et Li

Des prises de 1 000 g d'échantillons moulus ont été pesées
puis étalées dans des nacelles rectangulaires de quartz; ces
dernières ont été introduites dans un four à plasma d'oxygène
(LTA-600 Tracerlab) puis calcinées à basse température sous une
puissance de 260 W, selon FABRY (1971). La durée de calcination

était d'environ 12 h pour le blé et d'environ 38 h pour la pomme de terre. Les cendres obtenues ont été mises en solution dans 1 à 2 ml de HCl 30 % (Suprapur Merck No 318) par chauffage sur plaque de verre (Ceran Schott). Les solutions obtenues ont été filtrées sur un petit tampon de coton puis amenées à 50 ml (dosages de Fe, Mn et Zn) ou à 20 ml (dosages du Cu et Li). Les dosages des éléments Fe, Mn, Zn et Cu ont été réalisés par spectrophotométrie d'absorption atomique (SAA) selon PERKIN-ELMER (1976) avec un spectrophotomètre Perkin-Elmer 306, en flamme acétylène + air. Le Li a été dosé à l'aide du même instrument, mais par émission de flamme; en effet, cette technique s'est avérée bien plus sensible que la SAA pour le dosage du lithium.

METHODE DE DOSAGE DES ELEMENTS P, K, Ca et Mg

Des prises de 1 000 g d'échantillons moulus ont été pesées dans des capsules de platine, précalcinées sur une rampe de chauffage électrique Gerhardt, puis calcinées au four à moufle Gallenkamp (8 h à 520°C). Les pommes de terre, qui produisent des cendres noires, ont été reprises par quelques gouttes d'eau séchées à 100°C puis placées au four à 520°C durant 2 h. Les cendres obtenues ont été dissoutes dans 10 ml de HCl 16 %; on a évaporé à sec sur bain de sable, remis en solution dans 5 ml HNO_3 10 %, puis filtré sur papier filtre et amené à 100 ml dans des ballons jaugés.

Les solutions obtenues ont servi pour les dosages colorimétriques du P par formation du complexe jaune phosphovanadomolybdate (spectrophotomètre Beckman 24) ainsi que pour les dosages de K, Ca et Mg par SAA (PERKIN-ELMER, 1976).

METHODE DE DOSAGE DU B.

Des prises de 2 000 g d'échantillons ont été pesées dans des capsules de platine, additionnées de 200 mg de CaO, précalcinées sur rampe de chauffage puis placées au four à moufle durant 3 h à 450°C. Les cendres ont été mises en solution par 10 ml de H_2SO_4 1 N. Après centrifugation, le bore a été dosé dans la solution obtenue par réaction colorimétrique avec l'Azométhine-H (Sigma No A-3144) d'après SIPPOLA ET ERVIO (1977).

REPETABILITE ET EXACTITUTDE DES ANALYSES

Nous avons étudié la répétabilité des dosages d'Oligo-éléments en analysant 7 à 8 fois le même échantillon de blé moulu. Les résultats obtenus sont donnés dans le tableau 2.

Pour contrôler l'exactitude des dosages, nous avons analysés une farine de blé (Standard Reference Material 1567) provenant du National Bureau of Standards (U.S. Department of Commerce). Les résultats trouvés et les résultats certifiés sont indiqués ci-dessous (mg d'élément/kg matière sèche) :

	Trouvés	Certifiés
Fe	18,4	18,3 ± 1,0
Mn	8,5	8,5 ± 0,5
Cu	2,2	2,0 ± 0,3
Zn	11,3	10,6 ± 1,0
Ca	200	190 ±10
K	1380	1360 ±40

RESULTATS

Les résultats des analyses des grains du blé et des tubercules
de pomme de terre, ainsi que les rendements en grains à 85 % de
matière sèche et en tubercules frais sont donnés dans les tableaux 3
à 7. A partir des teneurs et des rendements, nous avons calculé les
exportations en oligo-éléments et en P, K, Ca et Mg par les produits
récoltés (tableaux 8 à 11).

DISCUSSION

Par comparaison avec les témoins,on constate que
l'intensification des cultures n'a pratiquement pas modifié les
teneurs en oligo-éléments (Fe, Mn, Cu, Zn, Li et B) dans les grains
de blé et dans les tubercules de pommes de terre. Les teneurs en
éléments minéraux mesurées dans les grains de blé cultivé à Changins
sont du même ordre de grandeur que celles obtenues en France par
MOREL, LASNIER et BOURGEOIS (1984) dans leurs essais de
fertilisation de longue durée à la Station agronomique de Grignon
avec la variété Champlein. Dans le cas des pommes de terre
cultivées à Changins et à Fey, l'intensification conduit à une
certaine diminution des teneurs en potassium et en magnésium des
tubercules. Cette observation devra être vérifiée par des essais
ultérieurs.

BIBLIOGRAPHIE

FABRY J., 1971. Etude de la calcination à basse température
            appliquée à la récupération d'éléments métalliques
            dans les matières végétales et animales, Thèse.
            Faculté des Sciences Agronomiques de l'Etat à
            Gembloux (Belgique).
MOREL R., LASNIER T. et BOURGEOIS S., 1984. Les essais de
            fertilisation de longue durée de la station
            agronomique de Grignon. INRA, Paris, 335 pp.
PERKIN-ELMER, 1976. Analytical Methods for Atomic Absorption
            Spectrophotometry. Norwalkm Connecticut, U.S.A.
SIPPOLA J. et ERRVIO R., 1977. Determination of Boron in Soils
            and Plants by the Azomethin-H Methods. Finn.Chem.
            Lett.,, 138-140.

Tableau 1

Caractéristiques des lieux d'essais et conditions culturales

	Changins parcelle 19	Fey parcelle 1B	Fey parcelle 6B
Culture	Pomme de terre	Blé	Pomme de terre
Précipitations moyennes (mm/an)	996	1150	
Températures moyennes (°C)			
de janvier	0,9	-1,0	
de juillet	19,0	16,7	
Insolations moyennes (h)	1692	1660	
Géologie	moraine	alluvions	
Nature des sols	moyen à lourd	moyen	léger à moyen
pH des sols	7,3	7,4	6,7
Argile (%) [0 – 2 μm]	30	16	14
Silt (%) [2 – 50 μm]	34	43	18
Matière organique (%)	2,6	2,1	2,3
N minéral 0 à 90 cm (kg N/ha)	32	25	32
Fumure de base:			
P_2O_5 (kg/ha)	80	70	110
K_2O (kg/ha)	240	120	300
Fumier (t/ha)	40	0	0

Tableau 2

Etude de la répétabilité des dosages d'oligo-éléments dans
un échantillon de grains de blé moulus

(Résultats en mg d'éléments/kg de matière sèche)

Elément	Nombre d'analyses	Minimum	Maximum	Moyenne	Ecart-type	Coefficient de variation (%)
Fe	7	37,3	41,8	39,8	1,39	3,5
Mn	7	49,5	53,3	50,8	1,22	2,4
Zn	7	31,1	33,5	32,1	0,97	3,0
Cu	8	3,7	4,1	3,8	0,14	3,7
Li	8	0,025	0,028	0,026	0,0014	5,4

Tableau 7

Teneurs en oligo- et macroéléments des grains séchés du blé
d'automne et rendements en grains (15 % H_2O)

Essai de culture intensive du blé d'automne
à Fey. Variété Arina

	0 kg N/ha	150 kg N/ha	150 kg N/ha + CCC + 2 fongicides	190 kg N/ha + CCC + 2 fongicides	190 kg N/ha + CCC + 2 fongicides + $MgSO_4$	230 kg N/ha + CCC + 2 fongicides	Moyenne
Fe (mg/kg)	40	37	39	34	32	42	37,3
Mn "	51	58	53	55	48	54	53,2
Zn "	32	27	29	26	25	36	29,2
Cu "	3,8	4,0	4,0	4,1	4,0	3,9	4,0
Li "	0,026	0,028	0,027	0,027	0,024	0,028	0,027
B "	1,4	1,2	1,8	1,2	1,1	0,9	1,3
P (‰)	4,21	4,18	4,18	4,14	4,13	4,36	4,20
K (‰)	4,55	4,47	4,38	4,59	4,62	4,46	4,51
Ca (‰)	0,52	0,48	0,46	0,44	0,47	0,50	0,48
Mg (‰)	1,20	1,19	1,14	1,14	1,14	1,14	1,16
Rendement (t/ha)	2,96	5,78	6,09	6,70	6,55	7,45	5,92

Tableau 4

Teneurs en oligo- et macroéléments des grains séchés du blé
d'automne et rendements en grains (15 % H_2O)
Essai de culture intensive du blé d'automne à Fey
Variété Bernina

	0 kg N/ha	150 kg N/ha	150 kg N/ha + CCC + 2 fongicides	190 kg N/ha + CCC + 2 fongicides	190 kg N/ha + CCC + 2 fongicides + $MgSO_4$	230 kg N/ha + CCC + 2 fongicides	Moyenne
Fe (mg/kg)	43	40	42	39	38	38	40,0
Mn "	53	55	59	55	59	55	56,0
Zn "	28	24	26	24	25	25	25,3
Cu "	4,1	3,6	3,6	3,6	3,4	3,6	3,6
Li "	0,025	0,025	0,024	0,022	0,023	0,023	0,024
B "	1,4	1,3	1,9	1,4	1,0	1,2	1,4
P (‰)	4,16	4,05	3,78	3,82	3,87	4,02	3,95
K (‰)	4,51	4,55	4,66	4,53	4,64	4,42	4,55
Ca (‰)	0,44	0,43	0,38	0,38	0,38	0,41	0,40
Mg (‰)	1,29	1,16	1,03	1,06	1,05	1,08	1,11
Rendement (t/ha)	3,00	5,45	6,19	7,17	7,22	7,49	6,09

Tableau 5

Teneurs en oligo- et macroéléments des grains séchés du blé
d'automne et rendements en grains (15 % H_2O)
Essai de culture intensive du blé d'automne à Fey
Variété Zénith

	0 kg N/ha	150 kg N/ha	150 kg N/ha + CCC + 2 fongicides	190 kg N/ha + CCC + 2 fongicides	190 kg N/ha + CCC + 2 fongicides + $MgSO_4$	230 kg N/ha + CCC + 2 fongicides	Moyenne
Fe (mg/kg)	37	39	35	34	36	35	36,0
Mn "	54	59	51	53	56	54	54,5
Zn "	29	27	25	24	26	27	26,3
Cu "	3,6	3,5	3,7	3,6	3,7	3,9	3,7
Li "	0,025	0,023	0,026	0,025	0,025	0,024	0,025
B "	0,9	1,3	1,0	1,2	1,3	1,1	1,1
P (‰)	4,15	4,03	3,82	3,91	3,96	3,90	3,96
K (‰)	4,36	4,48	4,60	4,58	4,48	4,20	4,45
Ca (‰)	0,46	0,41	0,39	0,39	0,39	0,40	0,41
Mg (‰)	1,18	1,16	1,01	1,02	1,02	1,01	1,07
Rendement (t/ha)	2,39	4,84	5,33	5,70	5,87	5,85	5,00

Tableau 6

Teneurs en oligo- et macroéléments des tubercules séchés de
pomme de terre et rendements des tubercules (poids frais)
Essai de fumure azotée sur la pomme de terre à Changins
Variété Bintje

	0 kg N/ha + fumier	60 kg N/ha + fumier	120 kg N/ha + fumier	180 kg N/ha + fumier	240 kg N/ha + fumier	Moyenne
Fe (mg/kg)	41	47	45	38	44	43,0
Mn "	6,1	5,7	6,9	7,1	6,4	6,4
Zn "	16	14	16	16	16	15,6
Cu "	6,0	5,9	6,2	6,3	6,2	6,1
Li "	0,051	0,046	0,045	0,047	0,048	0,047
B "	5,7	5,7	5,2	5,5	5,3	5,5
P (%)	0,31	0,27	0,27	0,24	0,25	0,27
K (%)	2,17	1,98	1,96	1,76	1,82	1,94
Ca (‰)	0,50	0,44	0,37	0,48	0,55	0,47
Mg (‰)	1,08	1,00	1,04	1,10	1,03	1,05
Matière sèche (% dans la mat. fr.)	23,7	25,5	24,2	24,9	24,4	24,5
Rendement (t/ha)	33,90	47,98	62,28	66,88	67,78	55,76

Tableau 7

Teneurs en oligo- et macroéléments des tubercules séchés
de pomme de terre et rendements des tubercules (poids frais)
Essai de fumure azotée sur la pomme de terre à Fey
Variété Bintje

	0 kg N/ha	60 kg N/ha	120 kg N/ha	180 kg N/ha	240 kg N/ha	Moyenne
Fe (mg/kg)	38	46	40	38	38	40,0
Mn "	7,8	10	8,8	10	14	10,1
Zn "	17	17	18	18	22	18,4
Cu "	6,8	6,6	6,4	6,6	7,9	6,9
Li "	0,044	0,044	0,043	0,042	0,040	0,043
B "	6,3	5,8	5,5	5,7	6,3	5,9
P (%)	0,26	0,28	0,25	0,25	0,24	0,26
K (%)	1,91	1,91	1,71	1,62	1,42	1,71
Ca (‰)	0,34	0,37	0,40	0,42	0,37	0,38
Mg (‰)	0,92	0,96	0,98	0,92	0,85	0,93
Matière sèche (% dans la mat. fr.)	25,6	25,4	25,4	24,5	23,8	24,94
Rendement (t/ha)	24,76	27,40	31,28	31,88	46,62	32,39

Tableau 8

Exportations d'oligo-éléments par les grains du blé d'automne (g/ha)

	0 kg N/ha	150 kg N/ha	150 kg N/ha + CCC + 2 fongicides	190 kg N/ha + CCC + 2 fongicides	190 kg N/ha + CCC + 2 fongicides + Mg SO$_4$	230 kg N/ha + CCC + 2 fongicides	Moyennes
Arina							
Fe	100,6	181,8	201,9	193,6	178,2	266,0	187,0
Mn	128,3	285,0	274,4	313,2	267,3	342,0	268,4
Zn	80,5	132,7	150,1	148,1	139,2	228,0	146,4
Cu	9,6	19,7	20,7	23,3	22,3	24,7	20,1
Li	0,065	0,138	0,140	0,154	0,134	0,177	0,13
B	3,5	5,9	9,3	6,8	6,1	5,7	6,2
Bernina							
Fe	109,7	185,3	221,0	237,7	233,2	241,9	204,8
Mn	135,2	254,8	310,5	335,2	362,1	350,2	291,3
Zn	71,4	111,2	136,8	146,3	153,4	159,2	129,7
Cu	10,5	16,7	18,9	21,9	20,9	22,9	18,6
Li	0,064	0,116	0,126	0,134	0,141	0,146	0,12
B	3,6	6,0	10,0	8,5	6,1	7,6	7,0
zénith							
Fe	75,2	160,4	158,6	164,7	179,6	174,1	152,1
Mn	109,7	242,7	231,1	256,8	279,4	268,5	231,4
Zn	58,9	111,1	113,3	116,3	129,7	134,3	110,6
Cu	7,3	14,4	16,8	17,4	18,5	19,4	15,6
Li	0,051	0,095	0,118	0,121	0,125	0,119	0,10
B	1,8	5,3	4,5	5,8	6,5	5,5	4,9
Moyennes des 3 variétés							
Fe	95,2	175,8	193,8	198,7	197,0	227,3	181,3
Mn	124,4	260,8	272,0	301,7	302,9	320,2	263,7
Zn	70,3	118,3	133,4	136,9	140,8	173,8	128,9
Cu	9,1	16,9	18,8	20,9	20,6	22,3	18,1
Li	0,06	0,12	0,13	0,14	0,13	0,15	0,12
B	3,0	5,7	7,9	7,0	6,2	6,3	6,0

Tableau 9

Exportations de P, K, Ca et Mg par les grains du blé d'automne (kg/ha)

	0 kg N/ha	150 kg N/ha	150 kg N/ha + CCC +2 fongicides	190 kg N/ha + CCC +2 fongicides	190 kg N/ha + CCC +2 fongicides + Mg SO$_4$	230 kg N/ha + CCC +2 fongicides	Moyenne
Arina							
P	10,6	20,5	21,6	23,6	23,0	27,6	21,2
K	11,4	22,0	22,7	26,1	25,7	28,2	22,7
Ca	1,3	2,4	2,4	2,5	2,6	3,2	2,4
Mg	3,0	5,8	5,9	6,5	6,3	7,2	5,8
Bernina							
P	10,6	18,8	19,9	23,3	23,8	25,6	20,3
K	11,5	21,1	24,5	27,6	28,5	28,1	23,6
Ca	1,1	2,0	2,0	2,3	2,3	2,6	2,1
Mg	3,3	5,4	5,4	6,5	6,4	6,9	5,7
Zénith							
P	8,4	16,6	17,3	18,9	19,8	19,4	16,7
K	8,9	18,4	20,8	22,2	22,4	20,9	18,9
Ca	0,9	1,7	1,8	1,9	1,9	2,0	1,7
Mg	2,4	4,8	4,6	4,9	5,1	5,0	4,5·
Moyenne des 3 variétés							
P	9,9	18,6	19,6	21,9	22,2	24,2	19,4
K	10,6	20,5	22,7	25,3	25,5	25,7	21,7
Ca	1,1	2,0	2,1	2,2	2,3	2,6	2,1
Mg	2,9	5,3	5,3	6,0	5,9	6,4	5,3

Tableau 10
Exportations d'oligo-éléments par les tubercules de pommes de terre Bintje (g/ha)

	0 kg N/ha	60 kg N/ha	120 kg N/ha	180 kg N/ha	240 kg N/ha	Moyenne
Changins						
Fe	329	575	678	633	728	589
Mn	49	70	104	118	106	89
Zn	129	171	241	266	265	214
Cu	48	72	93	105	103	84
Li	0,41	0,56	0,68	0,78	0,79	0,64
B	46	70	78	92	88	75
Fey						
Fe	241	320	318	297	422	320
Mn	49	70	70	78	155	84
Zn	108	118	143	141	244	151
Cu	43	46	51	52	88	56
Li	0,28	0,31	0,34	0,33	0,44	0,34
B	40	40	44	45	70	48

Tableau 11

Exportations de P, K, Ca et Mg par les tubercules de
pomme de terre Bintje (kg/ha)

	0 kg N/ha	60 kg N/ha	120 kg N/ha	180 kg N/ha	240 kg N/ha	Moyenne
Chandrins						
P	24,9	33,0	40,7	40,0	41,3	36,0
K	174,3	242,3	295,4	293,1	301,0	261,2
Ca	4,0	5,4	5,6	8,0	9,1	6,4
Mg	8,7	12,2	15,7	18,3	17,0	14,4
Fey						
P	16,5	19,5	19,9	19,5	26,6	20,4
K	121,1	132,9	135,9	126,5	157,6	134,8
Ca	2,2	2,6	3,2	3,3	4,1	3,1
Mg	5,8	6,7	7,8	7,2	9,4	7,4

CONTENT OF IRON, MANGANESE, COPPER, ZINC,
BORON AND LITHIUM IN GRAINS OF AUTUMN WHEAT
AND IN POTATO TUBERS (PRELIMINARY RESULTS)

Mr. J.P. Quinche, Mr. A. Maillard and Mr. W. Reust,
Federal Agronomic Research Station at Changins, Nyon

SUMMARY

Average content of trace elements and of P, K, Ca and Mg (in dry matter) of wheat grains (Arina, Bernina and Zénith varieties; six variants of nitrogenous fertilizer; Fey) and of potato tubers (Bintje; five variants of nitrogenous fertilizer; Changins and Fey):

	Wheat	Potatoes
Fe (mg/kg)	37.8	41.5
Mn (mg/kg)	54.6	8.3
Zn (mg/kg)	26.9	17.0
Cu (mg/kg)	3.8	6.5
Li (mg/kg)	0.025	0.045
B (mg/kg)	1.3	5.7
P (°/oo)	4.0	2.7
K (°/oo)	4.5	18.3
Ca (°/oo)	0.43	0.43
Mg (°/oo)	1.11	0.99

Average uptake of elements from the soil by wheat grains (Arina, Bernina and Zénith varieties; six variants of nitrogenous fertilizer; Fey) and by potato tubers) Bintje; five variants of nitrogenous fertilizer; Changins and Fey):

	Wheat	Potatoes
Fe (g/ha)	184	455
Mn (g/ha)	267	87
Zn (g/ha)	130	183
Cu (g/ha)	18	70
Li (g/ha)	0.12	0.49
B (g/ha)	6.1	62
P (kg/ha)	19.4	28.2
K (kg/ha)	21.7	198
Ca (kg/ha)	2.1	4.8
Mg (kg/ha)	5.3	10.9

In comparison with the controls, we find that a higher rate of fertilization has little or no effect on the content of Fe, Mn, Cu, Zn, Li and B in wheat grains and potato tubers. On the ohter hand, with 240 kg N/ha, we find reductions in the K and Mg content of potato tubers of as much as 25.7 per cent and 7.6 per cent respectively at Fey; at Changins, with the variant 240 kg/ha and manure, the decreases in comparison with the control are 16.1 per cent for K and 4.6 per cent for Mg. The same experiments are being repeated in 1986.

Use of Micronutrients in the German Democratic Republic on the Basis of Computerized Programmes

By Professors W. Podlesak and O. Krause
Institute of Plant Nutrition of the Academy
of Agricultural Sciences of the German
Democratic Republic, Jena

It is the task of agriculture in the German Democratic Republic to ensure food for the population on the basis of domestic raw materials. The annual growth rates of crop production reached 1 grain unit (GE) per hectare for the average of the years 1981 to 1985. In 1986, the grass crop yield amounted to 49.1 grain units per ha, the average cereal yield to 46.4 dt/ha. For wheat, the average yield was 56 dt/ha. The goal for 1990 is a grass crop production of 50 to 52 GE/ha and an average grain yield of 45 to 47 dt/ha.

Fertilization with micronutrients is of increasing importance for the planned stabilization of further increased yield levels since with rising yield levels both the sensitivity of plants in regard to micronutrient deficiencies is increasing, and also their micronutrient requirements. Furthermore, it is important to note that about two thirds of the agricultural land in the German Democratic Republic consists of light to medium soils of diluvial origin which by nature have relative limited supplies of micronutrients. About one third of the grassland consists of boggy soils with high ptt values where Mn and Cu are frequently available to plants in insufficient amounts only.

The analysis of the micronutrient content of the light and medium soils is increasingly incorporated into the generally established and legally regulated soil analysis which is repeated every 3 to 5 years. Thus the number of micronutrient tests carried out by the Agrochemical Testing and Advisory Service has been rising in recent years and has reached about 50 000 in 1986. In 1986, the following number of analyses broken down by elements were carried out and the following requirements for micronutrient fertilization (evaluated on the basis of DS 79) were established :

Number of micronutrient tests in 1986	Annual micronutrient fertilization requirements on the basis of DS 79
18 000 for Cu	140 000 ha for B
16 000 for B	130 000 ha for Cu
6 000 for Mn	35 000 ha for Mn
5 000 for Mo	10 000 ha for Mo
4 000 for Zn	4 000 ha for Zn

As these data are indicating, boron and copper are of the greatest importance under the conditions of the German Democratic Republic, but also for manganese, molybdenum and zinc there are some needs of fertilization.

Since 1970, the fertilizer advisory services are using computer programmes for a rational and economic application of macro (N, P, K, Mg, Ca) as well as of micronutrients (B, Cu, Mn, Zn). These programmes are constantly improved. Currently the third programme generation is in use and a new generation is under preparation.

All programmes used hitherto were based on big computers and referred to recommended fertilizer doses for individual field plots. By the way of a central computer in the Agrochemical Testing and Advisory Service these data were made available to all agricultural enterprises. These programmes have been used for the fertilization of about 5.7 millions hectares per annum and thus served practically the entire agricultural area of the country. Recommended doses for micronutrients were calculated in combination with macronutrients, which was in line with the fact that industry and trade are offering mostly N and P fertilizers with micronutrient supplementation (see Table 1).

A special advantage of these combinations of macro and micronutrients is the relatively even distribution of micronutrients in the soil and their availability to plants over a longer period of time. Another advantage is related to the fact that the residual effect of this way of fertilization - as far as the recommended dosages are respected - is amounting to about 2 to 3 years for B, to 5 years for Cu and to 3 to 5 years for Mo and Zn. It is a disadvantage, on the other hand, that the actual deficiency of a specific micronutrient does not always coincide with the actual macronutrient deficiency and it may therefore be difficult to find the special combination of the two. The residual effect is of little importance, furthermore, in those cases where some crops later in the rotation do not require a specific micronutrient that

was applied in the first year. For instance, this would be the case
with an application of B to sugar beets which are followed by
two years of cereals which do not require much B. However, B and Mo
are easily washed out. Thus the data given in Table 1 indicate the
usefulness of the fertilizers applied when taking into account the
total application and exportation from the soil of the pure
nutrients.

The high amounts applied are due to the fact that
micronutrients are normally bound to organic and mineral complexes
in the soil which reduces their availability to plants
considerably. This is why the full effect of the fertilization can
only be ensured by applying several times the calculated exportation
from the soil. However, a comparison of the rates of application
with the calculated exportation from the soil for the soil
application of P and N fertilizers with micronutrient
supplementation indicates a need to search for more efficient
micronutrient fertilization methods.

Currently the agricultural enterprises in the German Democratic
Republic are being equipped with office and personel computers as
well as with the corresponding soft ware for a computer-based
management of production. In this context, a new generation of
fertilizer programmes is under elaboration for a decentralised use
on the basis of microcomputers. This includes an independent
programme for the calculation of recommendations for micronutrient
fertilization by field plots. These fertilizer programmes form part
of a complex computerised system for the steering of soil and
livestock resources. A data bank for each field plot in each
agricultural enterprise, based on microcomputers, will serve as a
common basis for all programmes (Kundler, 1986).

The data bank of each agricultural enterprise contains all
relevant information on each field plot, such as natural conditions
and its geological origin, agrochemical, soil structure,
phytosanitary data as well as information on the annually practiced
cultivation questions. All programmes can be based on these basic
data banks and results obtained on the basis of such data may also
serve as an assistance for other and new programmes. Beginning in
1987, the whole system will be tested in agricultural practice,
including a programme for the calculation of fertilizer
recommendations for B, Cu, Mn, Mo and ZN.

Graph 1 shows that such recommendations are only calculated if

- a crop will be grown which is higly responsive to
  micronutrients, i.e. a crop which will produce significantly
  higher yields if micronutrients are applied on soils with a
  low to average micronutrient content.

- a soil analysis on the micronutrient content of the soil has
  been made.

The programme will inevitably suggest no application of
micronutrients if their level in the soil is high. The same holds
true for those cases in which soils with a low to average
micronutrient content have received an application following the
previous soil analysis and when the average period of the residual
effect has not been exceeded. If the programme suggests the
application of micronutrients if then also suggests the method of
application and quantities differentiated by micronutrient per ha.

Basically there are two alternatives in the fertilizer
programmes :

Alternative 1 - Optimal

    For each field plot, for which an application of
    micronutrients promises any economic return, a dosage of
    application is indicated.

Alternative 2 - Priority field plots for fertilization

    These are those, field plots under 1 for which a high
    economic ration can be expected, usually the additional
    value obtained should exceed the extra cost by
    100 per cent or more. This alternative will be
    implemented if supplies of micronutrient fertilizers are
    insufficient, thus requiring the setting of priorities.

Even in case of micronutrient sensitive crop and with a
relatively low micronutrient status of the soil, further factors
will be influencing the decision on an eventual fertilization. The
decision as to whether or not a specific crop is micronutrient
sensitive, is based on pot and field experiments.

As Table 2 indicates with the example of soil Cu, there are
considerable differences in the reaction of the various crops, even
within the same species, e.g. between wheat, barley and oats,

on the one hand and rye, on the other, wheat is particularly
sensitive to Cu deficiency whereas rye, in spite of the fact that
its Cu exportation from the soil corresponds more or less to the one
for wheat, shows nearly no reaction at all in yields to Cu
deficiency.

For other micronutrients similar experiments have been carried
out. Table 3 presents the micronutrient intensive crops under the
conditions of the German Democratic Republic. These data also make
a distinction between highly micronutrient demanding crops (given
the code 2 in the fertilizer programme) and those with an average
demand, resp. average fertilizing effect (given the code 2).

Generally speaking, micronutrient intensive crops produce 10 to
15 per cent higher yields if they receive the corresponding
micronutrient fertilization on soils low in these micronutrients.
In special cases, this yield increase can be much higher still.
Crops with average micronutrient requirements produce, under the
same conditions, about 5 to 10 per cent higher yields.

The determination of the micronutrient content of soils in the
German Democratic Republic is based on internationally recognized
standard methods, i.e. for B the hot water extraction method of
Berger and Trug. For Cu the HNO_3 method of Westerhoff is used,
for Mn the sulphite pH_8 method of Schachtschabel, for Mo
extraction of the soil with an oxalate solution of $pH_{3.3}$ according
to Grigg and for Zn the method of Trierweller and Lindsay based on
extraction with a 0.01 M EDTA/1 M ammoniumbicarbonate solution, are
being used. The indicative values shown in Table 4 are applied.

It is aimed at values (BFK-Sollwex) which are located around
the mean values given. The indicative values (Grenzwerte) are
differentiated by soil types and in the case of Mn also depending on
the ptt. In the case of Cu, account is taken of its reduced
availability in soils rich in organic matter and with a low sorptive
capacity. All values in excess of the indicative values are
considered as high and not requiring any micronutrient
fertilization. Values below the indicative values are considered as
low micronutrient contents.

These indicative values are not differentiated by type of crop,
but are relating generally to micronutrient sensitive crops. These
crops are usually recommended for micronutrient fertilization in
cases of low soil contents. In the case of average soil supplies
and average micronutrient sensitivity of crops, there are the
following additional criteria for a decision on the application of
micronutrient fertilizers :

- The yield level.

There are three yield categories, i.e. low, medium and high.
In many cases a high yield level will call for micronutrient
fertilization.

- The position of the determined micronutrient content in the
soil in relation to average supplies. The relatively higher
actual contents are usually not calling for an application
and the opposite is true for the relatively lower figures
which are actually determined.

- Soil pH.

High soil pH values are normally enhancing a positive
decision on B and Mn application. A further factor is the
use of ammonium sulphate which tends to reduce the pH and to
mobilize Mn. In the case of low Mo contents, the
recommendation is often limited to liming in order
to increase the availability of Mo in cases of a sub-optimal
soil pH. High soil pH values generally enhance a decision in
favour of micronutrient application.

- The use of organic fertilizers, when applied in the same year
as the growning of the crop, reduces the need for
micronutrient application. This refers to manure and to B,
Cu and Zn.

The entire model for decision on micronutrient application is
summarized in Graph 2. The recommendations are always connected
with suggestions for the way of application. The following key is
used for the ways of fertilization :

    0 - No fertilization
    1 - Soil application (B, Cu, Mo, Zn)
    2 - Band application (B)
    3 - Foliar application (Cu, Mn)
    4 - Foliar application for Mn, not required in cases of
         application of ammonium sulphate
    5 - Seed treatment with Mo
    6 - Soil application of Mo, not required with liming
    7 - Calling for a plant analysis

Table 5 presents the various fertilization methods. The
various methods involve varying quantities of micronutrients and
much importance is given to cost saving application.

The traditional soil application of micronutrients is above all recommended in cases of low soil contents and when the crop is requiring micronutrients already at the early stages of development. In the meantime, cost saving methods have been developed also for some of these cases, e.g. :

- Band application of B to sugar beets or also a precision placement in plant rows with a liquid B fertilizer. Band application of B represents a saving of about 50 per cent in B quantities applied compared to the application of B superphosphate, and this without any reduction in the B effect. An early application is required in order to allow the B to reach the root system of the crop. Liquid B fertilizers can be applied in combination with the herbicide. Betanal with a special band specific equipment. Similar possibilities are also seen for application in potatoes and vegetables.

- Considerable savings can be made through the seed application of Mo for lucerne, red clover, rapeseed, vegetables and sugar beets. Mo is particularly suited for seed application due to its low exportation values (3 to 10 g/ha). Thus Mo will last throughought the vegetation period, but without any reduced germination of the seed. The seed treatment is carried out with a powder for rapeseed, usually in combination with a treatment with an insecticide-fungicide powder (Bercema-Oftenol), in the case of beets through peletting (coating), but for lucerne, red clover or spinach, by dipping. Also in the latter cases, a peletting (coating) method is planned for the future. Experiments carried out in the years 1983 to 1985 have shown that the seed treated with 7.5 g of Mo/ha for winter rape, with 15 g for beets and spinach as well as with 30 g for lucerne gave the same effect as a soil application of about 1 kg of Mo/ha. This favorable effect of seed treatment with Mo can be explained by the high mobility of molybdate in the soil which means that it disposes well around the seed in the soil and can then be easily absorbed by the roots. This same effect is also playing in favour of the band application of boron.

- Foliar application is also reducing the required supplies. This application is used particularly in thoses cases where the supply of the plants during the early stages of development is ensured whilst a deficit of micronutrients

arises in the subsequent phase of intensive growth. For
these purposes the classic micronutrient fertilizers such as
copper, manganese and zinc sulfate or ammonium and
natriummolybdates are progressively replaced by newly
developed chelates, particularly lignite sulfonates, in the
German Democratic Republic. These lignite sulfonates can
easily be combined with pesticides and can then be applied in
a flexible manner. They are absorbed rapidly by plants and
are therefore particularly useful for the treatment of
deficiencies which are becoming apparent only in the course
of the vegetation period.

The computer-based programme for the calculation of recommended
micronutrient applications presented here is providing the basis for
medium term plans, since it is also including the crop rotation in
addition to the available supplies in the soils and other factors.
In those cases were no definite recommendation is given, a further
control by the way of plant analysis is recommended during the
vegetation period. The latter is another computer-based complex
programme for the formulation of fertilizer recommendation.

The complex plant analysis presents a method of diagnosis and
control during the vegetation period as to whether the nutritional
status is sufficient for high yields. At the same time, this method
also provides a control on the effects obtained for earlier
fertilization. It cannot be limited to only one nutrient but needs
to involve all nutrients relevant for an optimal nutrition of the
plants. In this context not only the availability in absolute terms
of each nutrient is of importances but also their relative ratio to
one another. For each type of crop, a special range of essential
macro and micronutrients needs thus to be analysed and this is why
the system has been named "complex plant analysis". This method can
be applied at present to all crops shown in Table 6 which means that
the necessary indicative values on nutrient supplies have been
established on a reliable basis so far.

As indicated, N, P, K and Mg are always given priority for
testing (marked with two asterix). Ca is recommended for analysis
above all in the case of dicotyle plants with a high requirement of
Ca, to a lesser extent for grains. Regarding micronutrients, crops
are strictly grouped according to their micronutrient intensity.
Thus for instance B and Mg need not be analysed in cereals. The
following plant analyses are recommended, on the other hand :
B - in rapeseed, sugar beet and lucerne; Cu - in wheat, barley, oats
(but not in rye) and lucerne; Mn - in wheat, oats and beets;
Mo - in red clover and lucerne; Zn - in maize. Different parts of
the plant have been used in this analysis, depending on the crop in
question (see Table 7).

For cereals, perennial leguminous crops and grasses those parts of the plant indicated will be used for plant analysis, they have to be cut 2 cm above the ground, avoiding any soiling. For winter rape and for potatoes, the highest fully developed leaves on the plant have to be used. In the case of maize, for the first test fully developed leaves have to be cut in the middle of the plant and for the second test - at flowering stage - leaves immediately around the maize cob. For sugar and fodder beets, the most recent fully developed leaves are used, but after removing the stems. In general terms, the material for plant analysis has to be collected during the periods of intensive growth. Here the possibilities of correcting any deficiency are the better the earlier if is detected and remedied.

The Agrochemical Testing and Advisory Service is in charge of the chemical analysis, of its interpretation and of the formulation of concrete measures. Graph 3 shows an example of a computer printout, containing the identification of the sample had been taken (agricultural enterprise, number of the field plot, crop, data of the sample, stage of vegetation, plant part analysed) and the results of the analysis. The results are presented in the forms of figures and also graphically in order to facilitate a quick first conclusion. The nutritional status is classified according to three categories : insufficient, sufficient, excedental. This gives an immediate impression of the nutritional status for all nutrients. In each category, further relative notification is provided on the status of individual nutrients.

Also in the case of micronutrients the lowest limit is set by a level, below which a deficiency is becoming apparent and at which any additional supply is leading to an increase in yields. In case of deficiencies in the plant material a foliar application is recommended. In the case of Cu in winter wheat, the computer printout does not only show the Cu content but also the Cu : N ratio in dry matter, expressed in ppm of Cu : % N. The higher the N uptake of the plant the higher the Cu content needs to be because otherwise the high N uptake is not converted into higher yields. This is the reason why the Cu : N ratio needs to be recorded for winter wheat. According to investigations by Kraehmer in the Institute of Plant Nutrition of Jena a need for Cu application arises if the Cu : N ratio is L 1.4 in winter wheat. Investigations in this Institute have further shown that Cu deficiency is leading to pollen sterility and hence to a reduced number of kernels per ear (Kraehmer, 1986).

Also the ratio of competing ions or nutrients to each other is of importance, e.g. the ratios of N : K, N : P; K : Ca; Ca : Mg; P : Zn and Cu : Mn. The graphical presentation of the results of the plant analysis facilitates the observation of such ratios and leads to initial conclusions without any specifically fixed indicative ratio.

A special brochure for the users of the plant analysis results contains detailed instructions on the optimal data of foliar application, on the type of fertilizer to be used and on the concentration of solutions for spraying. In view of the small quantities involved the general conditions for correcting a suboptimal micronutrient status during the growing season are particularly promising.

<u>Summary</u>

The German Democratic Republic has developed a medium-term planning of micronutrient application by individual field plots for all micronutrient intensive crops on the basis of a computerised programme. This programme is supplemented by an operative and equally computerized programme, based on complex plant analysis. The latter permits the control and adjustment of the micronutrient supply of growing crops during the vegetation periods.

It is the primary goal of these programmes to ensure adequate supplies of micronutrients according to the specific requirements of crops, rather than to enrich soil in micronutrients. It is impossible for economic reasons to reach a generally sufficient level of micronutrients in all soils. Only in the case of Cu, the application of 12 kg/ha on lighter soils and of 24 kg/ha on heavier soils would be required to reach a medium level of Cu in all soils. Furthermore, a soil enrichment is only temporarily possible for elements such as B and Mo which are characterized by a high degree of mobility in the soil. Finally, the protection of the environment also suggests a careful and rational use of micronutrients.

Increasingly scarcer and more expensive raw materials also suggest the desirability of highly efficient and specific methods of application for micronutrients. In this context, good experiences have been made with the band application of B to beets and with the seed treatment of rapeseed, lucerne, beets and spinach with Mo. These methods offer considerably reduced application rates with equal effects. Foliar application of micronutrients leads into the same direction. This method permits to correct relatively rapidly those deficiencies which become apparent only in the course of the vegetation period. In this context mention ought to be made of newly developed micronutrient fertilizers such as lignite sulforate and a liquid B fertilizer which can be mixed without problem with pesticides. Precise instructions are being elaborated for the various methods of application.

Literature

Kraehmer, R. : Ermittlung der Ansprueche der Getreidearten an die kupferversorgung des Bodens anhand von Gefaessver-suchen.
In : Tag. - Ber. Akad. der Landwirtschaft. - Wiss. der DDR. - Berlin (1986), 248, S. 135-143.

Kundler, P. : Entwicklung der computergestuetzten Bodenfuehrung.
In : Feldwirtschaft 27 (1986) 8, S. 342-345.

Table 1

Average Annual Exportation of Micronutrients in a Crop Rotation
of 60 % Cereals and with a Yield Level of 60 Grain Units per
Hectare as well as Required Quantities of Micronutrients for
Soil Application

Micro-nutrient	Annual exportation (g/ha)	Macronutrient fertilizer with micronutrient supplementation	Required quantity in kg of pure nutrient per ha	Soil application X times the annual exportation
B :	150-200	B-Kalkammons. (0.2 % B) B-Superphos. (0.3 % B)	1.5-2.0	10 – 13
Cu:	60-100	Cu-Superphos. (1.0 % Cu)	5.0	50 – 80
Mo:	3- 10	Mo-Superphos. (0.2 % Mo)	0.6-1.0	150 – 300
Zn:	200-300	Zn-Superphos. (1.0 % Zn)	5.0	16 – 25
Mn:	300-700		(>40)[1]	(60 – 130)[1]

[1] Not practiced due to high soil fixation of Mn, only foliar application.

230

Table 2

Determination of Cu Requirements of the Different Cereals,
Compared to Other Crops; Results of Pot Experiments
(Kraehmer, 1985)

Crop	Harvested material considered	Relative yield (%) (100 = yield at sufficient Cu)	Order	Requirements
Lindseed (for fibers)	Straw	19	1	
Wheat	Grain	20	2	
Barley	Grain	50	3	
Oats	Grain	60	4	High
Lucerne	Dry matter	62	5	
Spinach	Dry matter	66	6	
Fodder beet	Beet dry matter	70	7	
Sugar beet	Beet dry matter	73	8	Medium
Red clover	Dry matter	85	9	
Potato	Tuber	89	10	
Turnip	Beet	90	11	
Rapeseed	Grain	93	12	Low
Rye	Grain	96	13	

Table 3

Micronutrient Intensive Crops

Crop	B	Cu	Mn	Mo	Zn
High require-ments (2)	Sugar beet Fodder beet Rapeseed Lucerne Fodder cabbage Sunflower	Wheat Barley Oats Sunflower Lucerne Carrots	Oats Wheat Sugar beets Fodder beets Peas Beans	Lucerne Red clover Fodder cabbage Cauliflower Spinach salad	Maize Linseed Beans
Average requi-rements (1)	Maize Potatoes Field bean Mustard Red clover	Maize Beets Grasses Red clover Field beans	Barley Rye Maize Potatoes Rape seed	Oats Beets Rapeseed Lupines	Beets Potatoes Field beans Red clover Lucerne

232

Table 4

Indicative Values for Micronutrients (Average Contents) for Different Types of Soil

Soil Type	B ppm	Cu ppm Org. matter ≤ 4 %	> 4 %	Mn ppm pH – Values			Mo-Soil value	Zn ppm
1 – Sand	0.15–0.25	1.5–3.5	2.0–4.5	≤ 4.9 / 2.	5.0–5.8 / 5–10	>5.8 / 10–20	6.4–7.0	1.0–2.5
2 – Loamy sand	0.20–0.30	1.5–3.5	2.0–4.5	≤ 4.9 / 2.4	5.0–5.8 / 5–10	>5.8 / 10–20	6.4–7.0	1.0–2.5
3 – Sandy loam	0.25–0.40	2.0–4.5		≤ 5.4 / 5.10	5.5–6.5 / 10–15	>6.5 / 15–25	6.8–7.8	1.5–3.0
4 – Loam and clay	0.35–0.60	4.0–8.0		≤ 5.9 / 10.15	6.0–6.8 / 20–30	>6.8 / 30–50	7.2–8.2	1.5–3.0
5 – Boggy soils	0.15–0.25	2.0–4.0		≤ 6.8 / 5.15	≥ 6.9 / 10–20			

Table 5

Micronutrient Fertilizers, Fertilizing Methods and Optimum Rates of Application

Micronutrient	Micronutrient fertilizer		Rate of application in kg of pure nutrient/ha
Soil application			
B	B-Superphosphate	(0.3 % B)	1.5 - 2.0
	Liquid B fertilizer (area spraying)	(3.7 % B)	1.5 - 2.0
	Liquid B fertilizer (band application	(3.7 % B)	0.7 - 1.0)
Cu	Cu-Superphosphate	(1.0 % Cu)	5.0
Mo	Mo-Superphosphate	(0.2 % Mo)	0.6 - 1.0
	Ammonium molybdate (seed treatment)	(54.4 % Mo)	0.015
Zn	Zn-Superphosphate	(1.0 % Zn)	5.0
Foliar application			
B	Borax	(11.3 % B)	0.4 - 0.6
	Liquid B ferlizer	(3.7 % B)	0.3 - 0.5
Cu	Copper lignite sulforate	(5.0 % Cu)	0.4 - 0.5
Mn	Manganese-sulphate	(27.0 % Mn)	1.0
Mo	Ammonium molybdate	(54.4 % Mo)	0.3
Zn	Zinc sulphate	(22.8 % Zn)	0.3 - 0.5

Table 6

Crops and Tests Recommended for Complex Plant Analysis

Crop	Tests recommended									
	N	P	K	Ca	Mg	B	Cu	Mn	Mo	Zn
Winter wheat	XX	XX	XX	X	XX	–	XX	XX	–	X
Winter barley	XX	XX	XX	X	XX	–	XX	X	–	X
Winter rye	XX	XX	XX	X	XX	–	–	X	–	X
Oats	XX	XX	XX	X	XX	–	XX	XX	–	X
Brewery barley	XX	XX	XX	X	XX	–	XX	X	–	X
Spring barley	XX	XX	XX	X	XX	–	XX	X	–	X
Spring wheat	XX	XX	XX	X	XX	–	XX	XX	–	X
Maize	XX	XX	XX	XX	XX	X	X	X	–	XX
Winter rapeseed	XX	XX	XX	XX	XX	XX	–	X	X	–
Potatoes	XX	XX	XX	XX	XX	X	–	X	–	X
Sugar beets	XX	XX	XX	XX	XX	XX	X	XX	X	X
Fodder beets	XX	XX	XX	XX	XX	XX	X	XX	X	X
Red clover	XX	XX	XX	XX	XX	X	X	X	XX	X
Lucerne	XX	XX	XX	XX	XX	XX	XX	X	XX	X
Grasses	XX	XX	XX	XX	XX	–	X	X	–	–

XX = Strongly recommended.
X = Not strongly recommended.
– = Not recommended.

<u>Table 7</u>

<u>Crops, Dates of Samples to be Taken and Parts of the Plants
to be Used for Control of Nutrient Supplies in Growing Crops</u>

Crop	Date of sample taken	Part ot the plant to be analysed
Cereals	End of stocking till end of shooting phase FEEKES 4 to 10	Entire plant above the ground
Maize	Crop 40 to 60 cm high	Leaves at the middle of plants
	Shooting of flowers	Leaves at the middle of plants
	Flowering	Leaves attached to cobs
Winter rapeseed	Bud till full flowering stage	Recently fully developed leaves
Potatoes	Bud stage till forming of tubers	Recently fully developed leaves without the stems
Sugar beets	End June till end August	Recently fully developed leaves without the stems
Fodder beets	End June till end July	Recently fully developed leaves without the stems
Lucerne	Bud stage till flowering stage	Entire plant above the ground
Grasses	Begin of flowering (first growth)	Entire plant above the ground

Structure of the Micronutrient Programme

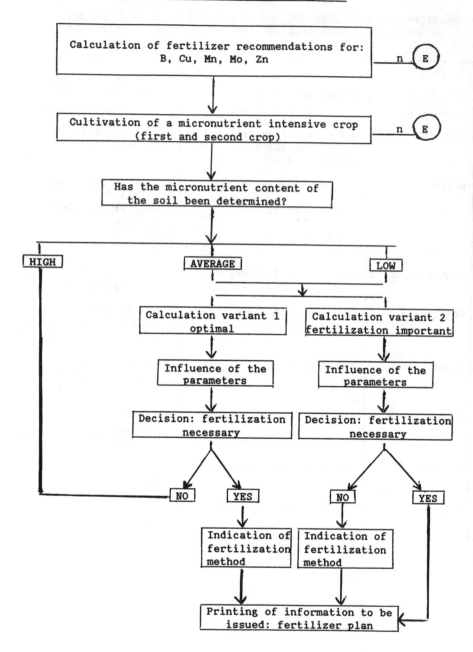

Abbildung 2:

Grundschema der Berechnung von Mikronaehrstoffduengungsempfehlungen

Variante 1 = optimal   Bedingungen fuer Empfehlung: "d u e n g e n"

niedriger Gehalt: hoher oder mittlerer Bedarf

mittlerer Gehalt: hoher Bedarf + hoher Ertrag
                               hoher Bedarf + mittlerer Ertrag + Istwert
                               < BKF
                               mittlerer Bedarf + hoher Ertrag + Istwert
                               < BKF
                               hoher Bedarf + mittlerer Ertrag + Istwert
                               > BKF + zusaetzl. Fakt.
                               mittlerer Bedarf + hoher Ertrag + Istwert
                               > BKF + zusaetzl. Fakt.

Variante 2 = vordringlich zu duengende Schlaege

niedriger Gehalt: hoher Bedarf + hoher oder mittlerer Ertrag
                               mittlerer Bedarf + hoher Ertrag

mittlere Gehalt: hoher Bedarf + hoher Ertrag + Istwert < BKF

Allgemeine Voraussetzungen fuer Empfehlungsberechnung:

mikronaehrstoffintensive Kultur, Vorlage eines Bodenuntersuchungs-
ergebnisses, Keine Mikronaehrstoffduengungsnachwirkung.

INSTITUT FUER PFLANZENERNAEHRUNG JENA   BEREICH AGROCHEMISCHE UNTERSUCHUNG UND BERATUNG

ERGEBNISSE DER PFLANZENANALYSE

BETRIEB: RAVENSBERG   KREIS: BAD DOBERAN   BEZIRK: ROSTOCK

PROBECHARAKTERISTIK	ELEMENT	GEHALT IN TR-MASSE	AUSR. BEREICH	VERGLEICHSDIAGRAMM (GEHALT) UNZUREICHEND : AUSREICHEND : UEBERSCHUSS
SCHLAG-NR : 163-0-0	N (%)	3.17	(2.40 - 4.80)	N :=================== : ===* :
PROBE-NR. : 10	P (%)	.35	(.31 - .54)	P :================= : =* :
FRUCHTART : W-GERSTE	K (%)	3.77	(3.40 - 4.80)	K :================ : ==* :
PRO.-TERMIN : 7. 5.	CA(%)	.64	(.30 - 1.00)	CA:=============== : =====* :
VEG.STADIUM : FEEKES 5/6	MG(%)	.13	(.08 - .19)	MG:================= : =====* :
PRO.ORGAN : GANZE PFL.	CU(MG/KG)	5.70	(5.00 - 12.00)	CU:============== : * :
	MN(MG/KG)	37.20	(30.00 - 150.00)	MN:============= : * :
SCHLAG-NR : 243-0-0	N (%)	2.78	(2.40 - 4.80)	N :============ : =* :
PROBE-NR. : 11	P (%)	.41	(.31 - .54)	P :=================== : ====* :
FRUCHTART : W-GERSTE	K (%)	3.67	(3.40 - 4.80)	K :=============== : =* :
PRO.-TERMIN : 7. 5.	CA(%)	.55	(.30 - 1.00)	CA:============= : ===* :
VEG.STADIUM : FEEKES 5/6	MG(%)	.14	(.08 - .19)	MG:================== : =====* :
PRO.ORGAN : GANZE PFL.	CU(MG/KG)	4.80	(5.00 - 12.00)	CU:===============* : * :
	MN(MG/KG)	22.70	(30.00 - 150.00)	MN:=========* : * :
SCHLAG-NR : 162-0-0	N (%)	3.74	(2.00 - 4.00)	N :================= : ========* :
PROBE-NR. : 12	P (%)	.42	(.24 - .61)	P :================ : =====* :
FRUCHTART : W-ROGGEN	K (%)	3.98	(2.80 - 4.30)	K :================= : =========* :
PRO.-TERMIN : 15. 5.	CA(%)	.39	(.32 - .95)	CA:============== : * :
VEG.STADIUM : FEEKES 7/8	MG(%)	.11	(.08 - .20)	MG:================ : ==* :
PRO.ORGAN : GANZE PFL.	CU(MG/KG)	6.40	(4.00 - 11.00)	CU:================= : ===* :
	MN(MG/KG)	37.00	(20.00 - 100.00)	MN:=============== : ==* :

FERTILIZER RECOMMENDATIONS FOR MAGNESIUM,
SODIUM AND SOME TRACE ELEMENTS IN RELATION
TO SOIL ANALYSIS IN THE NETHERLANDS

Dr. C.H. Henkens [1]/
Central Advisory Service for Soil, Water
and Fertilizer Management for Arable Crops
and Horticulture, Wageningen, Netherlands

I. INTRODUCTION

Unlike macronutrient deficiencies, micronutrient deficiencies
are often ascribed to disease. Originally copper deficiency was
mentionned as a disease related to land reclamation, manganese
deficiency was described as a grey speck disease of oats, and boron
deficiency in beets as heart rot. This confusion may have been due
to the fact that the various symptoms were known before the fact
that they actually resulted from any nutrient deficiency. Thus the
reclamation disease was known since 1855 in the Netherlands. First
in 1925 Hudig and Meyer cured this syndrome by application of copper
sulphate. The actual evidence for this disease being caused by
copper deficiency was not given before in 1938 by E.G. Mulderm
however.

Although magnesium is a macronutrient, it was related in so far
to trace elements in the Netherlands as Mg, deficiency in cereals
was described as "Hooghalen disease" and in potatoes as "Noorderling
syndrome" before it became known as an Mg deficiency.

The requirements of grass regarding Mg, Na and trace elements
are so small that no yield losses due to deficiencies of such
nutrients have been observed, at least in the Netherlands. These
nutrients are important for _animal_ health and production, however.
It is for this reason that the content of soils and grass of these
nutrients is subject of grassland research. For arable land,
attention is focussing on crops production (yields) only. Therefore
grassland and arable land are considered separately in this paper
which will review the fertilizer recommendation for Mg, Na and trace
elements.

II. GRASSLAND

Soil samples on grassland are taken to a depth of 5 cm in the
Netherlands. Fertilizer recommendations on the basis of soil
analyses are given for cobalt, copper, magnesium and sodium. The

---

[1]/ The paper was presented at the Symposium by Dr. T. Breimer.

manganese content of the grass depends on the soil pH. Other trace elements are of no importance in the Netherlands as yet.

Cobalt

Co is an essential constituent of vitamin B 12 substances. Rumen bacteria can use ingested Co for synthesis of vitamin B 12 that meets the ruminant's needs. So Co deficiency in cattle is in fact a vitamin B 12 deficiency.

In some countries, the recommended minimum content of Co in the ration is 0.07 mg/kg. It is difficult to prove that this content is critical because of the inherent problems of sampling feed or other forage free from contaminating soil. According to Piper (1944) soils contain 20–100 times as much Co as plant material. It is difficult to clean samples from the smallest soil particles which happens to be the very fraction with a high Co content. Therefore Co values in crops cannot be estimated in a reliable manner. It is also a question whether the Co content of grass is a good criterion for the Co supply of the animals. The animals take up soil as well and the soil Co can be utilised by the rumen bacteria. Co values in soils are considered to be a more reliable criterion than Co values in grass. Therefore in the Netherlands soils are analysed for Co by extraction with 2.5 per cent acetic acid.

Hart and Deys (1951) compared the soil on farms with Co deficiency in cattle and on farms with healthy cattle. He concluded that the cobalt content extracted with acetic acid should be 0.20–0.30 mg/kg.

Co values on sandy soils are usually low (Henkens 1959) Table 1 gives the recommendation for Co application to grassland based on the soil content of Co extracted with acetic acid (2.5 per cent).

Table 1

Cobalt Recommendation on Grassland

Soil Co content (mg/kg) (0.5 cm layer)	Recommendation (kg Co/ha)
< 0.1	0.5
0.1 - 0.3	0.3
> 0.3	0.0

Source : Anonymus 1986 (b).

Field trials have shown that a supply of 0.5 kg Co lasts for a period of 5 to 10 years (Table 2).

Table 2

Increase of Co Content in the Soil Extracted with
2.5 % Acetic Acid in the 0-5 cm layer by Supplying
0.5 kg in 1949

Soil type	Increase in Co value (mg/kg)				
	1949	1950	1951	1952	1959
Humic clay-soil	0.72	0.52	0.71	0.83	
Humic sandy-soil	0.44	0.39	0.54	0.39	
Very humic sandy-oil	0.60	0.59	0.63	0.47	0.11
Peat soil	0.68	0.56	0.51	0.65	0.20

Source : Henkens 1959 (a).

Copper

Copper is important for animal health and production. The absorbed amount of Cu is determined by the Cu content of the feed and its availability from the diet. An important factor inhibiting Cu absorption in the Netherlands is the concentration of sulphide in the rumen (Anonymous 1973).

Although a high Cu content in the grass does not guarantee a sufficient absorption of Cu by cattle it is advisable to prevent lower Cu contents. For this reason the relationship between the Cu content of soil (0.5 cm) and grass has been investigated. The Cu content in the soil is estimated by extraction with diluted nitric acid (0.43 mol HNO_3/litre). The relationship between the Cu content of the soil and of perennial ryegrass (Lolium perenne L) was studied in a pot experiment with 70 different humic sandy soils. One year before the experiment was started an amount of copper sulphate corresponding to 50 kg per ha was applied to a number of pots of each soil type. Figure 1 shows the relationship between soil Cu and the Cu content in grass.

Experiments on permanent grassland gave similar results as indicated by figure 2

There is almost no further increase in the Cu content of grass
after a soil Cu value of 5 ppm is reached. Application of copper
fertilizers to pasture soils containing 5 ppm of HNO_3 extractable
Cu or more is therefore of little use. It is generally accepted
that there is no risk for Cu deficiency in cattle as a consequence
of deficient intake when the Cu content of the grass reaches 6-7
mg/kg dry matter. Figures 1 and 2 show that this content is reached
if the Cu-HNO_3 value in the soil amounts ± 4 ppm

The results mentioned have led to a Cu fertilizer
recommendation based on soil analysis as mentioned in Table 3.

Table 3

Copper Recommendation on Grassland

Soil Cu (mg/kg) (0.5 cm layer)	Recommendation (kg Cu/ha)
< 2.0	6
2.0 - 4.9	3.5
≥ 5.0	0.0

Source : Anonymus 1986 (b).

Although the Cu content of grass is somewhat lower at lower pH
(pH-KCL < 4.0) and higher pH values (pH-KCL > 6.0), this is not
taken in to account. No more account is taken of the organic matter
content, although the Cu content of the grass is somewhat higher at
organic matter contents below 4 per cent (van Luit and Henkens,
1967). The copper dressings provide a sufficient Cu status of the
soil for approximately four years.

Magnesium

Mg is only applied on grassland with a view to animal health.
Mg absorption of cattle depends on the content of Mg, K and crude
protein of the ration. Kemp/Rameau (1962) have shown that the
higher the product of K x crude protein, the higher is the Mg
content needed in the grass. Figure 3 shows Mg requirements in
grass as a function of the content of K and crude protein.

Figure 3 shows that at K = 2.5 per cent and crude protein = 20 per cent (K x cp = 50) the Mg per cent needed is > 0.17 per cent in order to get a normal Mg content in the blood serum (> 2 mg Mg/100 cc blood serum). On fields with grass containing K x cp = 50, a sufficient Mg content can be reached by ferlizing with magnesium (Sluysmans 1967).

Grass with K x cp = 50-90 rarely contains sufficient Mg without a magnesium dressing. A supply of 100 kg Mg O/ha increased the magnesium content of the grass to a level which allowed for a magnesium content in blood serum of > 1 mg Mg/100 cc. Continous Mg fertilisation during 3 years gave magnesium contents in the grass on sandy soils which were sufficient for normal Mg levels in the blood serum (Prummel 1968).

Figure 3 shows that a product K x cp of > 90 demands a Mg content in the grass of > 0.25 per cent. Although continued high amounts of Mg after some years can give Mg contents in the grass of 0.25 per cent and higher, this is not recommended because of a corresponding decrease of the Calcium content (Sluysmans 1967, Kemp 1970, Dilz 1970).

These results have led to the magnesium fertilizer recommendations given in Table 4.

Table 4

Magnesium Recommendation on Grassland on Sandy and Loess Soils

MgO content in the 0-5 cm layer (mg/kg) extracted with 0.5 N NaCl	Recommendation (kg Mg O/ha)
< 75	200
75/150	100
151/250	50
> 250	0

Source : Anonymous 1986 (b).

This guideline for Mg dressing refers to sandy and loess soils and aims at:

1. Achieving and maintaining the Mg status of the soil
    ($\pm$ 150 mg Mg O/kg).

2. Achieving a sufficient Mg content in the grass in order to get a sufficient Mg supply to the animal except for the typical grass tetany periods.

To prevent grass tetany in spring and autumn direct supply of extra magnesium to the animal is necessary, either by foliar dusting with magnesium oxide or by providing magnesium rich cakes or pallets to the animals.

The Mg status of marine clay soils and peaty soils is generally higher. On these soils no fertilizer recommendation based on soil analysis is available because there was almost no correlation between soil Mg and the Mg content of grass. Furthermore on marine clay soils the effect of Mg dressings on the Mg content of grass is very small. To prevent grass tetany the supply of extra magnesium to cattle will be necessary.

Manganese

Manganese deficiency in cattle is not known in the Netherlands. Investigations on the relationship between soil Mn and grass Mn makes little sense since the Mn content of grass depends particularly on the soil pH. Figure 4 shows the relationship between pH-Kcl and the Mn content in grass on marine soils. On sandy soils the relationship is similar, but more scattered.

Sodium

Sodium deficiency in cattle can arise when the Na content in the feed does not meet the Na requirements of the animal. To meet the requirements of highly productive cows, an Na content of 0.15 per cent in the feed is sufficient (Anonymous 1973). The Na supply in winter time is usually sufficient as minerals are cummonly given and mineral mixtures generally contain sodium chloride. During the grazing period, however, the Na supply of the cattle depends particularly on the Na content of herbages.

The Na content of grass is largely determined by the Na content and the K number of the soil (Henkensen v. Luit 1963; Henkens 1965, a). The Na content of the soil is the amount of Na_2O soluble in 0.1 N HCL (mg/100 g soil) and the K number is the agricultural interpretation of K-HCL (Mg K_2 O/100 g soil extracted

with 0.1 N HCL) in which the empirically assessed influence
of organic matter is taken into account. At a given Na content
of the soil the Na content of grass decreases with increasing
K number of the soil. This decrease in Na content is small
at high K number (> 30). The Na content of the grass
increases with increasing contents of Na extracted with
0.1 N HCL. The influence of the sodium and potassium status
of the soil can be expressed on sandy soils by the ratio
$\frac{15(\text{K-number})}{\text{Na}_2\text{O}+6}$ (Figure 5) and on clay soils by $\frac{25(\text{K-number})}{\text{Na}_2\text{O}+14}$
(Figure 6) in which the numerator at K-number > 30 is the same as
that at K-number 30. With an increasing ratio the Na content of the
grass decreases.

The soil Na content may be used as a basis for Na fertilizer
recommendations, since with the potassium fertilizer policy in the
Netherlands, the K-number has generally reached a level at which it
has virtually a constant effect on the Na content of grass.

Table 5

Sodium Recommendation on Grassland for Sandy Soils

Soil Na$_2$O content (0-5 cm) in mg/100 g extracted with 0.1 N HCL	Recommendation (kg Na$_2$O/ha)	
	K-number $\leq$ 30	K-number > 30
< 2	60	80
2- 4	45	65
5- 9	20	40
10-13	0	20
> 13	0	0

Source : Anonymous 1986 (b).

Tables 5 and 6 show the recommendations on sandy respectively
clay and loess soils.

Table 6

Sodium Recommendation on Grassland for Clay and Loess Soils

Na$_2$ O content (0-5 cm layer) in mg/100 g extracted with 0.1 N HCL	Recommendation (kg Na$_2$O/ha)	
	K-number $\leq$ 25	K-number > 25
< 5	25	40
5- 7	10	25
8-10	0	10
> 10	0	0

Source : Anonymous 1986 (b).

The sodium dressing should be repeated every year.

III. ARABLE LAND

In the Netherlands fertilizer recommendations on the basis of soil tests are given for magnesium and the trace elements boron and copper. Sodium fertilizer recommendations to beets and recommendations to prevent or to cure manganese and molybdenum deficiency are also given, but these recommendations are not based on soil analysis.

Boron

Boron deficiency occurs in the Netherlands on sandy soils, where it is fairly widespread, on loess soils and - to a lesser extent - on river clay soils. Lehr and Henkens (1959) collected soil samples from beet fields in the Netherlands, both from B deficient plots and from plots which showed no signs of such deficiency. The soils were analysed for B by extraction with hot water and determination with curcumin (Truog method). In 1955, when the weather was warm and dry, heart rot occurred very frequently on sandy soils with a B content of less than 0.35 ppm. In 1956 when the weather was much milder and the incidence of heart rot was limited, soil B content of 0.30 ppm was established as critical. As a threshold value for use in advisory services for sandy and loamy soils 0.35 ppm B has been adopted.

To assess fertilizer recommendations pot and fields experiments were carried out in 1965-1968 (van Luit and Smilde 1969). Figure 7 shows that no B deficiency occured at a soil B content of > 0.35-0.40. There were no yield responses to B applications at soil B contents of 0.30-0.35 ppm and an application of 16 kg of borax (1.6 kg B) per ha was sufficient to prevent yields losses at low B contents (Figure 8).

These results have led to B fertilizer recommendation shown in Table 7. Although beet was the test crop in the field trials the recommendations also hold for alfalfa, swedes. maize, cellery and carrots which are as sensitive to B deficiency as beets.

Table 7

Boron Recommendation for Beets, Alfalfa, Maize, Swedes, Celery and Carrots

Soil B content (mg/kg) extracted with hot water	Recommendation (kg B/ha)
< 0.20	1.5
0.20-0.29	1.0
0.30-0.35	0.5
> 0.35	0.0

Source : Anonymus 1986 (b).

Copper

Copper deficiency may cause yield losses in cereals the most sensitive crop, already without any visual deficiency symptoms. Therefore soil testing for Cu is necessary. At the Institute of Soil Fertility in Haren oil Cu values extracted with 1 per cent EDTA and 0.43 N nitric acid have been investigated with cereals. Although the efficiency and the analytical errors of the EDTA and nitric acid methods are practically the same, the latter is preferable because it is faster (Henkens 1961). The method was tested with wheat and oats as test crops in a pot experiment with 70 different sandy soils and reclaimed peat soils. Furthermore a big number of field trials were laid out.

Figure 9 shows the relationship between soil Cu and the relative yield of spring wheat (variety Pako) in the pot experiments. This figure shows that at Cu values of 4 and above Cu application to wheat is not necessary.

Results of field trials were similar (Figure 10). Figure 11 shows the results for oats (variety Marne) in pot experiments. It shows that the soil Cu content for oats should be 3 and above, i.e. oats is less Cu sensitive than wheat. Judging from the responses of grain yields to Cu application, sensitivity decreases in the order wheat, oats, rye. Barley occupies an intermediate position between wheat and oats. Rye proved to be little sensitive to Cu deficiency. Considerable differences in sensitivity were found between the various wheat, oats and barley strains (Smilde and Henkens 1967).

Other crops are less sensitive to Cu deficiency than wheat. While 6 kg Cu/ha generally are sufficient for wheat and other crops are less sensitive, Cu rates as given in Table 8 are recommended. Results of van Luit (1975) show that the threshold values on sandy soils are valuable for other soils (e.g. marine clay and loamy soils) as well.

Table 8

Copper Recommendation on Arable Land

Soil Cu content (mg/kg) extracted with nitric acid	Recommendation (Kg Cu/ha)
< 3.0	6
3.0/3.9	2.5
≥ 4.0	0

Source : Anonymus 1986 (a).

Magnesium

The magnesium content in soils is determined by extraction with a solution of 0.5 N NaCl and expressed in Mg MgO/kg soil. Sluysmans (1961) investigated the response to ample magnesium application in relation to soil Mg. Figure 12 shows that Mg application has no effect on potato yields at a soil Mg of 80 Mg/kg and above.

Table 9 shows that beets, like potatoes, do not require any Mg application at a soil Mg of 80 mg/kg and above. For oats this value was established at 40 mg/kg.

Table 9

Yield Increase of Beet and Oats (kg/ha) in Response to Ample Mg Dressing at Different Soil Mg Contents Extracted with 0.5 N NaCl

MgO in mg/kg soil	10	20	40	60	80
Beets	–	2500	1700	900	100
Oats	1100	300	100	0	0

Source : (Sluysmans 1961).

For fertilizer recommendations, a soil Mg target of 45 mg/kg was chosen for cereal rotations and of 75 mg/kg for other rotations (Table 10).

Table 10

Magnesium Recommendation for Arable Land on All Soils Except Clay Soils

Cereal rotation kg MgO/ha	Other rotation kg MgO/ha
(45–soil magnesium) (weight of arable layer)	(75–soil magnesium) (weight of arable layer)

Source : Anonymus 1986 (a).

The recommendation refers to magnesium/sulphate as fertilizer. For magnesium/carbonate applied in autumn 50 per cent and applied in spring 25 per cent of total Mg applied is effective. The residual effect of magnesium/carbonate is better however.

On clay soils no Mg fertilizer application is recommended. In case of symptoms of Mg deficiency spraying with magnesium/sulphate is recommended.

Manganese

Manganese deficiency occurs in the Netherlands on light marine
soils where it is fairly wide spread, and on sandy and loess soils
with a higher pH. de Groot (1956) collected soil samples from
fields in the Netherlands, both from Mn deficient plots and from
plots which showed no signs of such deficiency.

Soils were analysed for Mn by extraction with a neutral
solution of normal ammonium acetate, containing 0.2 per cent
hydroquinone as a reducing agent. In Figure 13 the soil Mn of the
plots with and without Mn deficiency is related with pH-H_2O. This
figure shows that on sandy soils Mn deficiency is not found at
pH-H_2O of 6.0 and below. Between pH-H_2O 6.0 and 6.8 deficient
as well as healthy plots are found. All sandy soils with a pH-H_2O
of > 6.8 are deficient. This figure also shows that the soil Mn on
healthy, and deficient sandy soils is below 60 ppm. Marine soils,
which have a high pH by nature, are deficient at soil Mn of below
60, but deficiency may also occur at higher levels.

On the marine soils Mn is correlated with the organic matter
content. With organic matter contents of < 2 1/2 per cent crops
deficient with a soil Mn of < 60 ppm and healthy with a soil Mn of
> 70 ppm. At higher organic matter contents soils Mn should be at
least 100 ppm although deficiency may still occur (Figure 14). Thus
soil Mn values can be used to predict the occurence of Mn deficiency
on soils with a higher pH. It is not used as yet in the Netherlands
for fertilizer recommendation. Soil application of Mn prevents Mn
deficiency. However, high amounts (400-600 kg of manganese/sulfate)
are needed and the residual effect is negligible, therefore this
measure is too expensive (Henkens 1962). Therefore spraying with
1.5 per cent manganese/sulfate (1000 l/ha) is recommended instead to
cure Mn deficiency.

Investigations on the movement of Mn in plants (Henkens and
Jongman 1965, b) have shown that the Mn applied on leaves is
transported to roots, but that the redistribution from the roots is
insufficient to prevent Mn deficiency in the foliage which may
developp after spraying. In practice this means that if the soil Mn
supply is insufficient plants should be sprayed several times
throughout the season. This conclusion was confirmed by results
obtained in field trials with wheat, oats, beets and peas.

For wheat, oats and beets the first spraying is best applied as
soon as Mn deficiency symptoms become evident, and the second
spraying some four weeks later, at least when new leaves have been
formed. For the control of marsh spot of peas the spray should be
applied during flowering. It is best to do this in the middle of

the flowering season and to repeat the spraying at the end.
Spraying pea crops prior to flowering does not influence the
percentage of marsh spot (Henkens 1962).

Molybdenum

Molybdenum deficiency is not wide spread in the Netherlands.
Henkens (1972) conducted an investigation on the occurence of Mo
deficiency in beets in the Netherlands. In addition, the factors
influencing Mo uptake were studied by means of pot trials. A soil
may cause Mo deficiency symptoms in beet if it contains more than 1
or 2 per cent of iron (soluble in 10 per cent HCL). Whether such
deficiency did occur depended primarily on th pH, but particle size
and the form of the iron also played a rule. Since Mo deficiency
occurs ab pH-KCL < 5.4 it is generally recommended to prevent a
lower pH. To cure an Mo deficiency soil application of 2-3 kg
sodium or ammonium molybdate per ha is recommended. As Table 11
shows, spraying with 0.05 per cent sodium/molybdate (500 l/ha) also
gives good results.

Table 11

Effects of Fertilization and Spraying with Sodium/Molybdate on Beet and Sugar Yields

Rates	Beet yield (tons/ha)	Sugar yield (kg/ha)
Control	31.4	5151
3 kg sodium molybdate/ha	34.9	5688
Sprayings 500 l/ha :		
0.01 % sodium molybdate sol	32.8	5364
0.05 %	36.0	5427
0.10 %	35.0	5727

Source : Henkens 1957.

Sodium

Sodium is only recommended for beets on sandy soils at a rate
of 200 kg Na_2O per ha. Field trials have shown considerable yield
increases on these soils (Henkens 1971; vab Burgea 1983). As the Na
content of these soils is low by nature and the residual effect is
negligible, soils are not analysed for Na.

252

Literature

Anonymus 1973. Tracing and treating mineral disorders in dairy
    cattle.
    Centre for Agricultural Publishing and Documentation,
    Wageningen, NL.

Anonymus, 1986. a. Advies basis voor de bemesting van bouwland.
    Ministerie van Landbouw en Vissery. CAD voor Bodem,
    water en Bemestingszaken in de Akkerbouw en Tuinbouw.
    Wageningen, NL.

Anonymus, 1986. b. Advies basis voor de bemesting van grasland en
    voedergewassen. Ministerie van Landbouw en Vissery.
    CAD voor Bodem, water en Bemestingszaken in de
    Veehoudery. Wageningen, NL.

Bur van P.F.J, M.R.J. Holmes and K. Dilz, 1983. Nitrogen supply from
    fertilizers and manure on yield and quality of sugar
    beet.
    Symposium, Nitrogen and sugar beet, Bruxelles 1983
    (16-17 febr.).
    Intern. Inst. for sugar beet research, pp. 189-282.

Dilz, K. en G.H. Arnold, 1970. De magnesiumvoorziening van grasland.
    2. Verhoging van het Mg-gehalte van weide gras op
    zandgrasland door gebruik van magnesamon. Stikstof 6,
    150-157.

Groot, A.J. de, 1956. Influence of age and organic matter on the
    availability of manganese in marine and estuary soils.
    VIe Congrès Interne de la Science du Sol. Paris, 1956.

Groot, A.J. de, 1963. Manganese status of Dutch and German holocene
    deposits in relation to mud transport and soil
    genesis. Versl. Landbouwk. Onderzoekingen Nr 69.7.
    Pudoc, Wageningen, NL. 164 pp.

Hart, M.L. 't en W.B. Deys, 1951. Verslag C.I.L.O. over 1950,
    137-144.

Henkens, Ch.H., 1957. Onderzoek over molybdeengebrek.
    Landbouwvoorl. 14, 213-217.

Henkens, Ch.H., 1959. a. Kobalt op grasland. Landbouwvoorl. 16,
    642-651.

Henkens, Ch.H., 1961. The copper content of the soil determined with
          biological and chemical methods.
          Versl. Landbouwk. Onderzoekingen 67. 10.
          Pudoc, Wageningen, NL. 28 pp.

Henkens, Ch.H., 1962. Bedeutung des Kupfers für Ackerbau und
          Grünland.
          Landwirtsch. Forsch. 16, Sonderh. 56-65.

Henkens, Ch.H., 1962. Manganmangel und dessen Beseitigung.
          Landwirtsch. Forsch. 16, Sonderh. 66-71.

Henkens, Ch.H and B. van Luit, 1963. Determination of the sodium
          status of grassland with help of soil analysis.
          Versl. Landbouwk. Onderzoekingen 69-13.
          Pudoc, Wageningen, NL. 52 pp.

Henkens, Ch.H., 1965. a. Factors influencing the sodium.
          Content of meadowgrass.
          Neth. J. Agric. Sci. 13, 21-47.

Henkens, Ch.H. en E. Jongman, 1965. b. The movement of manganese in
          the plant and the practical consequences.
          Neth. J. Agroc. Sci. 13, 392-407.

Henkens, 1971. Natrium bemesting by bieten.
          Bedryfsontw. 2 (1971), 39-46.

Henkens, Ch.H., 1972. Molybdenum uptake by beets in Dutch soils.
          Agricultural Research Reports 775.
          Pudoc, Wageningen, NL. 54 pp.

Hudig, J. en C. Meyer, 1925. De ontginningsziekte en haar
          bestrijding.
          Bericht Rijksl. Bouwproefstation Groningen 18, 6 pp.

Kemp, A. en J.Th.L.B. Rameau, 1962. Voorstel betreffende de
          adviesgeving ter voorkoming van kopziekte, gebqseerd op
          de chemische samenstelling van grasmonsters die door
          veehouders worden ingezonden.
          Gestencild rapport I.B.S.

Kemp, A. en J.M. Geurink, 1970. Effects of magnesium dressing on the
          magnesium content of soil, grass and of blood serum
          from milking cows.
          Versl. Landbouwk. Onderzoekingen 747.
          Pudoc, Wageningen, NL. 33 pp.

Lehr J.J. and Ch.H. Henkens, 1959. Threshold values of boron
    contents in Dutch soils in relation to boron deficiency
    symptoms in beet (heart rot).
    Transactions of the World Congress of Agricultural
    Research, Rome.

Luit, B. van en Ch.H. Henkens, 1967. Invloed van de kopertoestand
    van de grond op het kopergehalte van gras en klaver.
    Versl. Landbk. Onderzoekingen 695.
    Pudoc, Wageningen, NL. 33 pp.

Luit, B. van en K.W. Smilde, 1969. Boriumbemesting van suikerbieten
    gebqseerd op grondonderzoek.
    Rapport 9, Instituut voor Bodemvruchtbaarheid, Haren,
    47 blz.

Luit, B. van, 1975. De kopertoestand van zeekleigronden.
    Rapport 1, Instituut voor Bodemvruchtbaarheid, Haren.

Mulder, E.G., 1938. Over de betekenis van koper voor de groei van
    planten en micro-organismen, in het bijzonder een
    onderzoek naar de oorzaak der ontginningsziekte.
    Diss. Wageningen, NL. 133 pp.

Piper, C.S., 1944. Soil and Plant Analysis, New York, 362 blz.

Prummel, J., 1968. Invloed van magnesium- en kalibemesting op het
    magnesiumgehalte van weidegras.
    De Buffer 14, 72-77.

Sluijsmans, C.M.J., 1959. Beziehungen zwischen Magnesiumgehalte des
    Bodems, Mangelsymptomen und dem Mehrertrag
    niederländischer Böden.
    Landw. Forschung 13, Sonderheft 17-23.

Sluijsmans, C.M.J., 1961. Werkwijze en resultaten van het
    magnesiumonderzoek.
    Landbouwvoorl. 18, 55-59.

Sluijsmans, C.M.J., 1964. Grondonderzoek als basis voor de bemesting
    van haver met magnesium.
    Versl. Landbk. Onderz. 643.
    Pudoc, Wageningen, NL. 37 pp.

Sluijsmans, C.M.J., 1967. Invloed van bemesting met kieseriet en
    kalizout op het magnesiumgehalte van weide gras.
    Gestenc. Versl. van interprovinciale proeven 120,
    P.A.W. Wageningen, NL.

Smilde, K.W. en Ch.H. Henkens, 1967. Sensitivity to copper
        deficiency of different cereals and strains of cereals.
        Neth. J. Agric. Sci. 15, 249-258.

Smilde, K.W., 1976. Minor elements in the nutrition of cereals.
        Compte rendu des séances; Séminaire d'étude
        céréaliculture, Gembloux, 1976, 303-312.

256

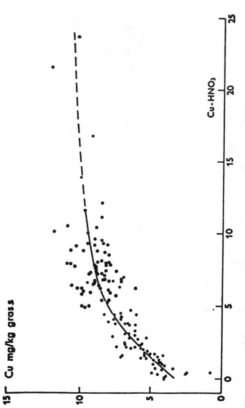

Fig. 1 . Relationship between soil copper content and copper content of perennial ryegrass after correction to equal humus content (4 %) and pH-KCl (4.2). (v. Luib and Henkens, 1967)

• Soil Cu before

o Soil Cu after

> application of copper sulphate

257

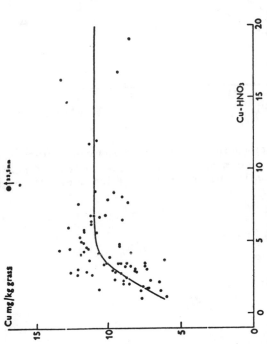

Fig. 2 . Relationship between soil copper content and copper content of grass after correction to equal nitrogen content (3 %). (v. Luik and Henkens, 1967)

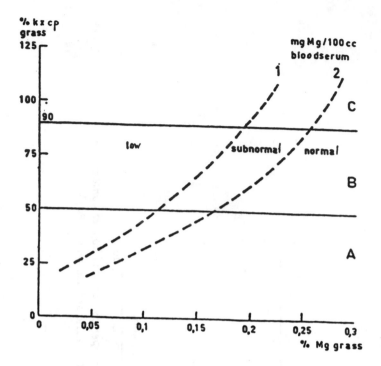

Fig. 3 Relationship between magnesium concentration in the
blood serum of cattle and the content of magnesium,
potassium and crude protein (cp = b.25N) in grass.

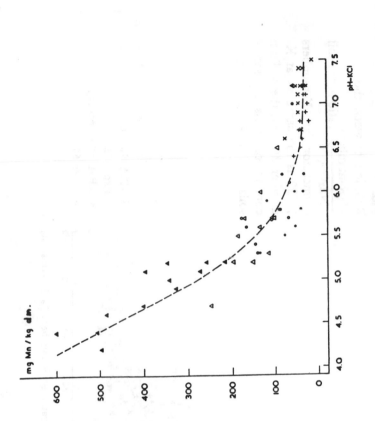

Fig. 4 Relationship between pH-KCl and manganese content of grass on marine clay soils (Henkens, lgbz)

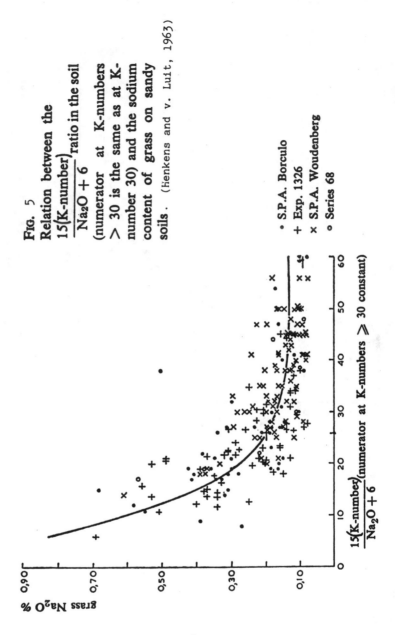

Fig. 5
Relation between the
$\dfrac{15(\text{K-number})}{\text{Na}_2\text{O} + 6}$ ratio in the soil
(numerator at K-numbers
> 30 is the same as at K-
number 30) and the sodium
content of grass on sandy
soils. (Henkens and v. Luit, 1963)

• S.P.A. Borculo
+ Exp. 1326
× S.P.A. Woudenberg
○ Series 68

$\dfrac{15(\text{K-number})}{\text{Na}_2\text{O} + 6}$ (numerator at K-numbers ⩾ 30 constant)

grass Na$_2$O %

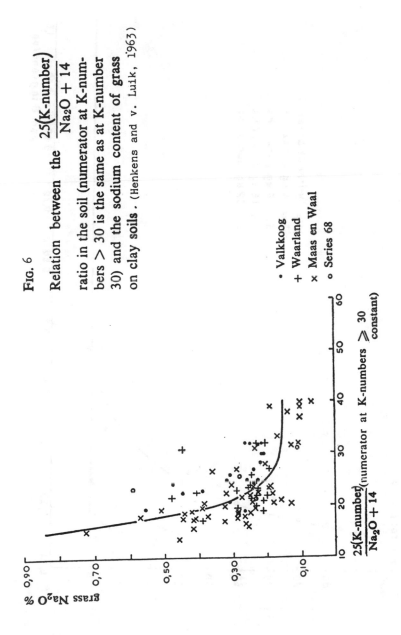

FIG. 6

Relation between the $\dfrac{25(\text{K-number})}{\text{Na}_2\text{O} + 14}$ ratio in the soil (numerator at K-numbers > 30 is the same as at K-number 30) and the sodium content of grass on clay soils. (Henkens and v. Luik, 1963)

• Valkkoog
+ Waarland
× Maas en Waal
○ Series 68

262

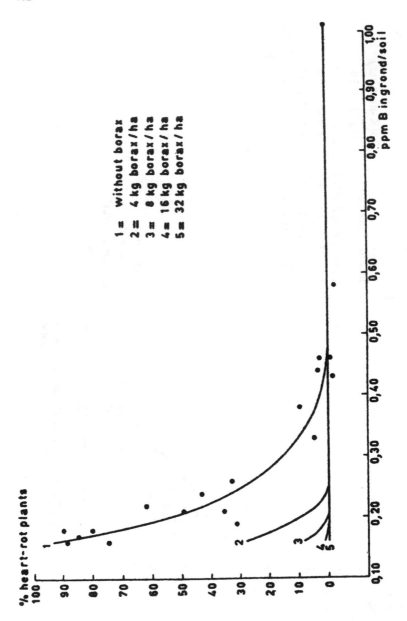

.Fig. 7 Relationship between soil boron and percentage of plants affected by heart-rot; field experiments. (v Luit and Smilde, 1969)

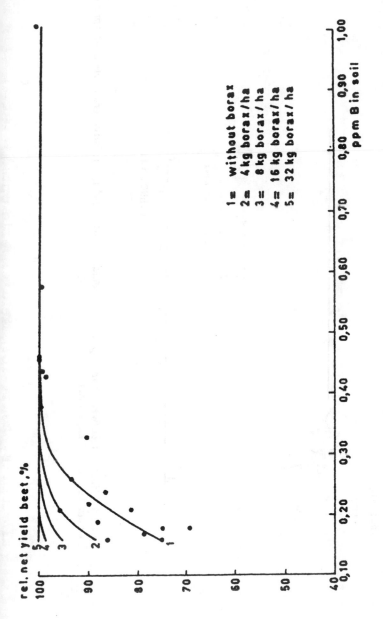

Fig. 8 Relationship between soil boron and relative beet yield (with 32 kg borax/ha=100%); field experiments. (v. Luik and Smilde, 1969)

264

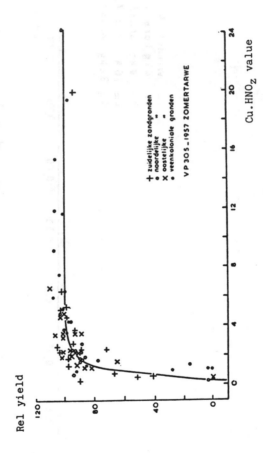

FIG. 9 *Relation between the copper nitric acid value and the relative grain yield of spring wheat in the pot experiment* (Henkens, 1961)

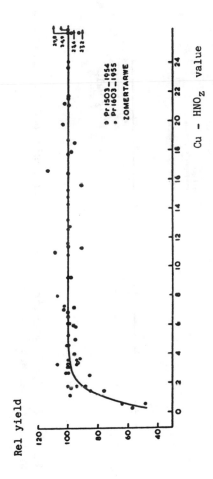

FIG. 10 *Relation between the copper nitric acid value and the relative grain yield of spring wheat (1954 and 1955)* (Henkens, 1961)

266

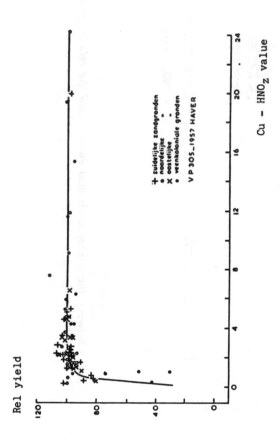

Rel yield

Cu — HNO₂ value

FIG. 11 . *Relation between the copper nitric acid number and the relative grain yield of oats in the pot experiment* . (Henkens, 1961)

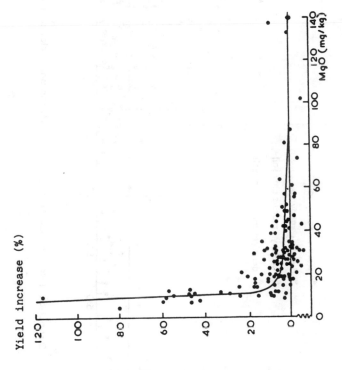

Fig. 12 Relationship between response of potatoes to magnesium dressings
and soil magnesium extracted with NaCl (1M) on sandy soils.
(Sluysmans, 1961)

268

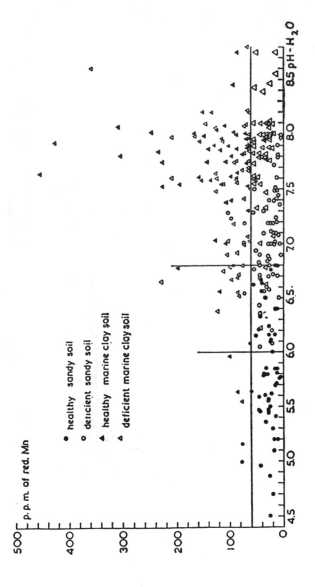

Fig. 13   Distinction of deficient and healthy diluvial sandy and marine clay soils with regard to Mn

(de Groot 1956, 1963)

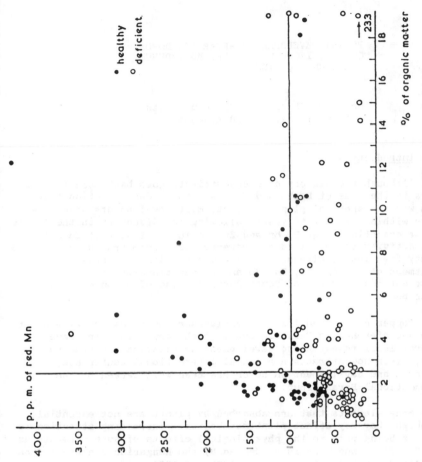

Fig. 14 Distinction of deficient and healthy marine soils with regard to Mn
(Beets and cereals) (de Groot, 1963)

RECENT DEVELOPMENTS REGARDING BORON,
COPPER, IRON, MANGANESE, MOLYBDENUM,
SELENIUM AND ZINC

Dr. V.M. Shorrocks, Micronutrients
Bureau, Tring, United Kingdom

1. Introduction

Although the history of micronutrients goes back more than 120
years to the time of Sachs and Knop and their demonstration that
iron was an essential plant nutrient, most developments have taken
place within the last 60 years following the discovery in the 1920's
of the essentiality of B, Mn and Zn and in the 1930's of Cu and Mo.
Most activity in the field has however been concentrated in the last
twenty-five years, although there are exceptions, notably the
treatment of Mn, Co, Cu, Zn, Mo and Fe deficiencies in the
"Micronutrient Desert" of South Australia and of boron deficiency on
sugar beet.

Copper celebrated its 50th anniversary as an essential element
with a conference in Perth in 1981; both copper and zinc have
international research and development associations devoted to
them; iron now attracts a biennial international conference;
research and development on boron has been well served by the
International Borax Group.

Many elements that are absorbed by plants are not essential,
although they may affect plant metabolism. Increasing attention is
in fact being paid to the physiological effects of such elements as
Ni, Se, Ti, V and W as is evidenced by the Hungarian conferences on
"Hardly Known Trace Elements" in 1985 and 1986.

It is quite likely that more elements will be found to be
capable of affecting plant growth, and some may even be shown to be
truly essential. It seems improbable however that any such new
elements will be found to be of widespread importance.

We should be careful however not to become complacent. Whilst
molybdenum and selenium are only required in extremely small
quantities by plants and animals respectively, they can nevertheless
be of great economic importance, as is clearly the case for
molybdenum in many parts of Australia and for selenium in parts of
China where supplementation of both the animal and human diet is
essential.

The object of this paper is to review recent developments of
significance with trace elements and crop production worldwide, to
make a few predictions and to indicate where research is needed.

The selection of items for highlighting is inevitably subjective, being based on the worldwide activities of the Micronutrient Bureau over the last six years.

Various items of veterinary and medical interest are included in order to emphasise that animals and man, at the end of the food chain, are the ultimate recipients of micronutrients and are applied to pastures and arable crops. It is becoming more readily appreciated that maintenance of good animal and human health should start with correct nutrition of crops and pastures, and that it should not be left to the fire brigade remedial approach of treating the severely deficient animal or human.

An attempt has been made in the Tables for each micronutrient to list those crops for which conventional wisdom declares them to be susceptible to the particular deficiency, together with the crops which are frequently treated worldwide.

2.   Boron

Boron has been historically associated with root crops since 1931 when BRANDENBURG first demonstrated the importance of boron on sugar beet in the field and the finding short after, in Scotland and Ireland, that swedes and turnips had a marked requirement for boron.

Crops Susceptible to Boron Deficiency

Alfalfa	Cotton	Pines
Apple	Eucalyptus	Red Beet
Broccoli	Grapes	Rutabaga
Cauliflower	Groundnuts	Sugar Beet
Carrot	Oil Palm	Sunflower
Celery	Oilseed Rape	Swede
Coffee	Olive	Turnip

It is curious that the two most significant recent developments concern two oilseed crops namely sunflower and oilseed rape.

2.1   Sunflower

The sensitivity of stem elongation of sunflowers to boron supply resulted in the plant being used in a diagnostic pot test to assess soil available boron many years ago, but it is only since sunflowers became an important crop in their own right that agronomists have studied the requirements for boron.

The breakthrough with sunflowers and boron deficiency came with the work of BLAMEY (1976) in South Africa; he was the first to realize and to emphasize that symptoms other than the classical ones on the growing point were of significance. The symptoms he recorded are now well known, including the development of necrotic areas on midstem and upper leaves, the various malformations of the flowering head, and also the breaking of the stem and the fall of the flower and seed head.

Boron deficiency symptoms have since been reported in Argentina, Bulgaria, France, Spain, Thailand and the USSR, and it is now evident that the crop should be classified with sugar beet and many of the brassicas as being highly susceptible to boron deficiency.

Developments are now awaited in other countries where sunflower is an important crop, such as Australia, China, Czechoslovakia, and Romania. There are no indications at present that boron is important on sunflowers in Canada and the USA.

2.2 Oilseed Rape

The story of boron deficiency on oilseed rape commenced in 1957 in Japan (KANNO) where transplanted rape was found to suffer from a "shrinking" disease; the crop failed to get away in the spring. Agronomic and physiological studies demonstrated that the problem was boron deficiency, and there the story remained for about 20 years, although there is some evidence that during this time there was limited use of boron on rape crops in Scandinavia. Expansion in the cultivation of oilseed rape following the breeding of improved varieties brought with it, particularly on the heathland soils of northern Germany and Denmark, the realisation that boron nutrition was important. Work in these countries, which led to use of boronated nitrogenous fertilizers (e.g. boronated ammonium sulphate nitrate) as well as the foliar application of boron, was followed by work in the United Kingdom where, despite the incidence of only mild symptoms, yield responses of 13-20% were obtained. At the same time, workers in the DDR had established that oilseed rape was a boron responsive crop (BRUCHLOS et al. 1979).

The explanation for the yield response in the absence of clearly defined symptoms is probably due to the particular requirement for boron in pollen development, for pollen tube growth and fertilization. The workers in Schleswig-Holstein demonstrated that the boron response was attributable to an increased number of seed per locule and not to any increase in seed weight (TEUTEBERG, 1978).

Substantial increases in yield following boron application have also recently been reported in Australia, China and the USA.

2.3 Future

The ease with which boron deficiency is prevented and controlled at relatively low cost means that there is little incentive for studies on genotypic efficiency in boron utilization, although there are indications from studies on sunflower and tomato of genotypic variation in boron absorption that could make this an interesting line to follow.

Situations that warrant further investigation with respect to boron include, in Europe, cereals, potatoes, grapes, and possibly clover/grass pastures, and in the Tropics, seed legumes, and groundnuts.

2.3.1 Cereals

Whilst cereals are certainly not susceptible to boron deficiency, the use of boron on cereals has been an established practice in Finland for many years and boronated fertilizers applying corrective doses of about 1 kg/ha boron have been commonly used. Responses to around 20% were reported on barley, oats and wheat (TAHTINEN, 1970). Boron had been used without phytotoxicity because it was not applied with the seed; boron toxicity readily occurs on cereals when farmers mistakenly combine drill seed with a boronated fertilizer.

There have been reported of boron deficiency increasing pollen sterility in cereals, e.g. work in Brazil (DA SILVA 1983) demonstrated that boron applications of 0.65 kg/ha boron increased wheat yields by reducing male sterility not only on the treated crops but also on the following wheat crop.

2.3.2 Potatoes

The situation regarding potatoes is rather curious. In the 1930's and 1940's responses to boron were obtained in Scotland, Austria, France and the USA. Since then interest in the boron nutrition of potatoes has waned although in the DDR potatoes are classified as a responsive crop (BORCHMANN et al. 1972). Despite these findings the conventional wisdom in western Europe is that potatoes are susceptible to boron toxicity. The explanation probably lies in the method of application; if the boron is in contact with the tubers shoot growth will be impaired.

2.3.3 Clovers

Whilst there is no doubt that seed formation in clovers is very sensitive to boron supply through its effect on pollination, there is relatively little information demonstrating benefits on clover in pastures - even on the maintenance of the clover content. There is however, an established use of boron in regenerating pastures and where coastal bermuda grass is grown under intensive management in the USA (SPOONER et al. 1983).

2.3.4 Grapes

It has been known for many years that grapes are capable of responding to boron applications, e.g. when growing on the steep eroded soils of the Moselle (GARTEL 1974). Despite the finding, in most grape growing countries, of the fairly characteristic symptom known as "hen and chickens" or "millerandage", there is much conservatism when considering boron use on grapes. The situation is unlikely to change in "quality" vineyards but where maximum production is the aim, it may be different.

3. Copper

The draining of heath and the moorland soils in Europe particularly in Denmark and Holland in the 1920s resulted in the first recorded cases of copper deficiency (HUDIG et al. 1926). Copper deficiency was commonly known as reclamation disease on oats, barley, rye, beet and leguminous crops; it was quickly found that the symptoms could be cured by the soil application of copper sulphate. The finding of copper deficiency on cereals on the coastal calcareous sand in South Australia in the late 1930s (PIPER 1938) marked the start of much work on the extensive areas of copper deficiency in soils in Australia, that has continued up to the present time. Recent reports in the national press there have indicated that copper deficiency is still limiting the area over which wheat can successfully be grown in Victoria.

The biocidal properties of copper had been exploited for just over 100 years and it was the finding, over 70 years ago, of benefits to plants following the application of copper fungicidal sprays to fruit in Florida in the absence of any pathogenic disease organisms (FLOYD 1913), that led some workers to suspect that copper was essential. It is curious that in the same year that SOMNER (1931) published the first unequivocal evidence of the essentiality of copper to plants, a field response on cattle was demonstrated (NEAL et al. 1931).

Table 2. Crops susceptible to copper deficiency
Crops which are often treated are underlined

Alfalfa	Lettuce	Spinach
Barley	Oats	Red Beet
Carrot	Onions	Tobacco
Citrus	Rice	Wheat

3.1 Cereals

Whilst it is widely appreciated that cereals are particularly susceptible to copper deficiency, it is only relatively recently with the work of ALLOWAY and co-workers (TILLS et al. 1981) and MORARD (1985) that the significance of sub-clinical copper deficiency on cereals is becoming acknowledged. The fascinating physiological and morphological studies carried out by these two researchers, working independently, has revealed the degeneration and imperfect development of both anthers and pollen in wheat suffering from sub-clinical copper deficiency that can substantially limit yields.

The question that now has to be answered is how widespread is sub-clinical copper deficiency on cereals both in Europe and elsewhere, especially on soils where clinical copper deficiency formerly occurred.

Many find it difficult to accept that it is worthwhile treating sub-clinical deficiencies, which makes it all the more important for this work to be carried to a further stage, either by means of plant analytical or soil analytical surveys to delineate more accurately the soils and conditions under which sub-clinical copper deficiency can be expected.

3.2 Sugar Beet

The finding by TILLS et al. (1981) that sugar beet, not showing any symptoms, could respond to copper, opens up yet another new field; the mechanism by which sugar beet responds must be entirely different from that in cereals.

3.3 Pastures and Grazing Animals

Attention continues to be focused on the copper nutrition of ruminants, and particularly into understanding the factors involved in affecting and controlling the absorption and utilization of copper by animals. It is widely accepted for example in the United Kingdom, that sub-clinical copper deficiency in ruminants is widespread, and it is therefore unfortunate that blood analysis for copper does not accurately allow the identification of animals that will necessarily develop clinical symptoms. Animals can be treated by oral dosing with copper sulphate, by mineral supplements in the feed, by copper containing licks, by the implacement of copper needles of glass boluses, by the addition of copper to drinking water and by copper injections. These direct methods of providing copper are normally preferred by veterinarians.

The grazing animal is of course reliant, under natural conditions on the forage for the major part of its copper supply, and it is therefore natural to consider administering copper via the pasture. Unfortunately this method has its limitations as it is difficult to elicit sufficient and long-lasting increases in the copper content of the herbage of pure grass swards, particularly of Italian ryegrass. Clover/grass swards are normally richer in copper, and it is easier to improve their copper content. However most of the work on pastures has involved soil surface applications which, because of the poor mobility of copper, do not increase the copper content in the rooting zone. Application and incorporation of copper at time of pasture re-seeding would improve the availability of copper to the grass roots, and long-term residual effects, as have been obtained with cereals, would be expected.

3.4 Mitigation of Decomposition of Peat

A novel and noteworthy agricultural use of copper is that demonstrated by MATHUR et al. (1979) in Canada, who have shown that the breakdown of peat can be reduced by about 50 per cent by the application of fairly heavy dressings of copper sulphate (60 kg/ha for 3 years). The explanation is that the enzymes exuded by the microbes that are normally responsible for the degradation of the large insoluble compounds in peat are complexed and inactivated by metals, of which copper is particularly active.

Having demonstrated the efficacy of recently applied copper, the Canadian workers have shown in a series of papers that there are likely to be no deleterious effects of the heavy copper applications on crops, soil or the environment e.g. LEVESQUE et al. (1983). It now remains to be seen how relevant this work is for cultivated peat soils in other countries such as the USA, Indonesia, Israel and Ireland.

There are already indications from short-term laboratory incubation studies in Ireland that copper applications can reduce the breakdown of a woody fen peat (MCDONNELL 1986).

3.5 Pig Slurry

Consideration of copper in agriculture cannot ignore the environmental aspects concerning the disposal of copper-rich pig slurry.

The inclusion of copper in the diet of pigs to improve feed conversion and promote growth results in pig slurries containing large amounts of copper e.g. around 1,000 ppm Cu on dry matter basis. Concern is naturally expressed regarding the safety aspects for animals particularly sheep, on grassland that has recently been treated with pig slurry and also on the long-term effects of enriching the soil with copper. Some recent work in Ireland (MCGRATH et al. 1982) has shown that whilst herbage is easily contaminated, the increase in plant copper, which occurs as a result of absorption of copper from the soil, is extremely small even following the cumulative application of slurry copper amounting to 49 kg/ha copper. The severity of the direct contamination of the vegetation depends not only on the rate of application but also on the sward density, the time interval and also on rainfall; good agricultural practice should ensure that grazing animals are not returned to the treated pasture too quickly and that opportunities for soil splash contamination of the herbage minimized.

Work in Virginia that has been in progress since 1978 continues to show that massive applications of copper in pig manure (totalling 250 kg/ha copper over 8 years) has not had any deleterious effect on maize yields on three soil types; in another study the annual application of copper sulphate (totalling 388 kg/ha copper zinc has not had any effect on maize yields (MARTENS et al. 1986).

3.6 Human Nutrition

Evidence is steadily accumulating that the low copper content of many western diets is implicated in, not only arthritis and inflammatory disorders, but also in lipid metabolism and cardiovascular disease.

There are now reasons for believing that whilst adults may actually need less copper than the recommended daily allowance of 2-3 mg/day copper (WHO), many people do not obtain sufficient copper.   For example, Swedish studies showed an average uptake of 1.6 mg Cu, in Denmark 1.7 mg Cu and in the United Kingdom 1.8 mg Cu (MAFF 1981).

In animal tissues, copper is necessary with zinc for the activity of the enzyme superoxide dismutase which, together with glutathione peroxidase (which contains selenium), is involved in providing a defence against superoxide radicals.   These radicals are produced in joints in the case of rheumatoid inflammations. Some good results have already been obtained with the separate administration of copper, zinc and selenium, and further work is required involving combined supplementation.

The hypothesis linking copper deficiency with coronary heart disease is largely based on evidence from copper deficient rats which show abnormal cardiac electro physiology, and from the fact that humans with coronary heart disease exhibit several characteristics that are also found in animals suffering from copper deficiency.

Recent disturbing results from an experiment in the United States of America on humans fed 1 mg copper per day that had to be terminated early because of heart abnormalities has further focussed attention on copper (REISER et al. 1985).

3.7 Future

The clear message coming not only from animal studies, but also from human dietary and physiological studies, is that grassland and many of our arable crops are of low/marginal copper status, not only in their own right but for the animals consuming them. The way ahead must surely be through improving the copper status of soils and crops generally. At the same time, it is recognized that the poor mobility of copper in the soil, in the plant as well as across the gut wall means that close attention needs to be paid to the efficiency of utilization.

4.   Iron

It is now about 140 years since it was first realized that
lime-induced chlorosis (on grapes in France) was due to iron
deficiency, and still it is widespread and agronomists still search
for efficient and economic ways of correcting the deficiency.

About 40 per cent of the world's soils are prone to iron
deficiency (VOSE, 1982). The fact that iron deficiency symptoms can
readily be seen in many countries where soils are alkaline or
contain free calcium carbonate, suggests that either farmers are not
aware of suitable treatments, that products are prohibitively
expensive, or that farmers do not believe that correcting iron
deficiency is worthwhile.

When chelates first appeared on the agricultural scene in the
early 1950s, it was believed that deficiencies not only of iron but
also of other trace elements would be satisfactorily corrected by
the soil application of chelates. Unhappily that has not proved to
be the case despite the publication of more than 1,300 papers on
chelates in plant nutrition. It is accepted that it is economically
viable to apply expensive chelates such as Fe-EDDHA, to high value
perennial fruit crops but that it is neither economic nor practical
to apply chelates to most arable crops, such as sorghum and soybeans
which are particularly sensitive to iron defiiency. Whilst foliar
sprays are effective in the short term, the very poor
re-translocation of iron from older to younger leaves necessitates
repeated application; any treatment or practice that improves the
phloem mobility of iron would be valuable.

Table 3. Crops susceptible to iron deficiency
Crops which are often treated are underlined

Citrus	Groundnuts	Soybeans
Field Beans	Mint	Sudan Grass
Flax	Ornamentals	Fruit Trees
Grapes	Sorghum	Vegetables.

4.1   Iron Stress Mechanism and Genotypic Variation

Great advances have been made in understanding the nature of
the iron deficiency problem and particularly of the plant's reaction
to an inadequate iron suppy by groups of workers in Germany (e.g.
ROMHELD, 1984), in the Netherlands (e.g. BIENFAIT, 1985) and the USA
(e.g. OLSEN, 1981).

Whilst much of the work may be thought to be highly academic,
there is little doubt that it has great relevance for the future
control of iron deficiency. It has already been found that many
crops, including chick peas, citrus, dry beans, field beans,
groundnut, jute, maize, oats, pigeon pea, rice, soybeans, sunflower,
and sugarcane exhibit a wide genotypic variation in their ability to

respond to iron stress. Some genotypes are iron efficient and are able, either by excluding reducing agents or hydrogen ions to solubilize iron in the vicinity of their roots. Some species, notably the grasses, have the ability to release ligands that chelate iron, but it is not yet known whether genotypes differ in their ability to produce such ligands.

There is much that now needs to be done to exploit this genotypic variation and particularly in the breeding of new varieties that combine iron efficiency with other desirable characteristics. This work will take many years, but already with soybeans, it has proved possible to breed iron efficient soybeans which, although not in the top yielding group, are able, when sown together with high yielding but iron inefficient soybeans, to contribute to yield by producing reasonable yields in those portions of the field where the iron inefficient soybean would fail (FEHR, 1982).

4.2 Siderophores

Microbially produced chelating agents, otherwise known as siderophores, continue to attract much attention, not only with regard to the part they play in mobilizing iron for their own use, and possibly for use by plants, but also because of the relevance, through the mechanism of microbial competition for iron, with the suppression of root fungal disease organisms that are sensitive to iron supply (e.g. ELAD, 1985).

Developments can be confidently expected concerning the use of microbially produced siderophores as potential iron sources and also in combating root fungal diseases.

4.3 Treatment of Iron Deficiency

Spray application of inorganic iron, of complexed iron, of iron chelates and the soil application of Fe-EDDHA and on some soils of less stable chelates, remain the preferred methods of treatment.

New treatments that are receiving attention include:

   (i) application of acidified iron-rich mining residues close to the roots.

  (ii) pressure injection of iron salts and iron chelates into trunks of trees.

 (iii) ploughing in of green manure of, for example, post harvest grain sorghum stubble or iron rich Amaranthus that had been sprayed with iron sulphate.

  (iv) acidification of soil by sulphur

   (v) use of iron complexes

5. Manganese

It has been appreciated for many years that cereals, potatoes, sugar beet and peas, commonly suffer from manganese deficiency, especially following liming. One of the earliest occurrences of manganese deficiency was on oats growing along limestone roads in South Australia (SAMUEL et al. 1928). In contrast manganese deficiency is rare on acid soils or when acidifying fertilizers are applied close to cereal seed.

As a result of the wide occurrence of the deficiency, managanese is somewhat of a "work horse" amongst trace elements and it is surprising that it is not as well catered for as other micronutrients, either with commercial technical service or with Research Associations.

Table 4. Crops susceptible to manganese deficiency
Crops which are often treated are underlined

Apple	Oats	Sorghum,
Barley	Peas	Spinach
Beans	Peaches	Strawberries
Citrus	Potato	Sudan Grass
Grapes	Radish	Sugar Beet
Lettuce	Soybean	Wheat

5.1 Soybeans

Much work has recently been carried out on the manganese nutrition of soybeans in the USA, and the results are similar to those reported on other crops. Foliar applications of relatively small amounts of manganese (1.12 kg/ha), if applied early, are as effective in increasing soybean yields as massive applications of inorganic manganese to the soil (30 kg/ha) (MASCAGNI et al. 1984, GETTIER, et al. 1985). Soil application of manganese chelates and manganese complexes have been found to be as effective as inorganic soluble manganese salts only when applied at the same rate of manganese.

The two groups of workers stress the importance of timing; manganese must be sprayed as soon as the symptoms appear.

Critical manganese levels in leaves and shoots have been assessed with a fair measure of agreement between the workers; deficient top mature leaves contain less than 20 ppm Mn.

Some measure of success has also been obtained, using a double acid (HCl and H_2SO_4) technique for assessing critical manganese levels in the soil. Not unexpectedly, it has been found that the critical level increased with soil pH, from about 2 mg/kg at pH 5 to 7.9 at pH 7.

5.2 Sweet Lupins

Work in Western Australia (HANNAM et al. 1984) has confirmed that the split seed disorder of sweet lupins is a manifestation of manganese deficiency and that it is genetically linked to the low alkaloid lupin cultivars. When the deficiency is severe, growth and yields are reduced and maturation is likely to be delayed. Where the deficiency is slight, few symptoms may be apparent during the vegetative stage, but the seed coats are liable to split exposing the cotyledons.

Work on methods of preventing manganese deficiency on lupins has followed that on soybeans in that the early remedial measures developed in Australia were based on the banded application of manganese sulphate at 15-30 kg/ha; this treatment was not always completely successful, and farmers were deterred by the relatively high cost. At the same time foliar applied manganese was not always satisfactory, and it was thought that the exact timing of the foliar application was important.

All too often, the timing of foliar sprays of micronutrients is a subject that is conveniently ignored. Most micronutrients are only weakly phloem mobile and it is very important therefore that foliar applications are timed to coincide with periods of maximum requirement.

Sweet lupins have a large requirement for manganese when grain is filling, which occurs almost simultaneously on all flowers. It is not surprising that HANNAM et al. (1985) found that the single foliar application of manganese sulphate at mid-flowering of the first order apical lateral inflorescences was the best treatment; manganese applied either earlier or later was less effective.

Unfortunately such timing means that if the foliar sprays were applied from conventional tractor mounted equipment, there would be considerable damage to the crop. Low volume sprays are fortunately effective; the single application of 0.7 kg/ha manganese in 20 litres via micronair atomizers prevents both foliar and seed symptoms.

5.3 Cereals and Disease

There are many examples where micronutrients have been implicated in disease incidence and severity, e.g. boron deficiency and ergot, and zinc deficiency and oidium on Hevea brasiliensis (BOLLE-JONES et al. 1957). Unfortunately in the majority of cases the trace element/disease study was incidental to the main investigation, and as a result there are many leads that could well be followed up in greater detail. One such study has been carried out in Australia involving take-all on wheat.

5.4 Take-all

The severity of take-all (Gaeumannomyces graminis) is known to be worse on soils of high pH and in cool, wet soils. Graham and Rovira (1984) have shown that the susceptibility of wheat to the disease depends in part on the availability of manganese in the soil, and is inversely correlated with the concentration of manganese in the host tissues. They artificially infected the soil and then added different amounts of manganese sulphate. The benefits of manganese were observed both on plants that were mildly deficient (but showing no symptoms), as well as on ones that were adequately supplied with manganese.

The mechanism whereby manganese reduces take-all has yet to be established and there are only indications that foliar applies manganese may be effective.

Wherever take-all is a problem and wheat may be of low manganese status, soil application of manganese is likely to be worthwhile.

5.5 Solubilisation of Manganese in the Rhizosphere

A major difficulty in predicting manganese deficiency is that soil analysis is generally unreliable due very probably to such factors as unsuitability of extractant, variations introduced by handling of samples and soil pH.

Studies on the manganese content in soil solutions extracted from the rhizosphere and non-rhizosphere regions of field grown barley plants have indicated that there is considerable solubilisation, not only of manganese but also of zinc, copper and cobalt in the rhizosphere as it develops (LINEHAN and et al. 1985). Whereas a poor correlation was found between the manganese content in barley shoots and the amounts of chemically extractable manganese in the soil, a good correlation was found to exist between manganese extracted by CaCl from fresh, moist "rhizosphere soil" and plant manganese content on sites where the manganese solubility was improved by the application of acidifying nitrogenous fertilizers.

The solubilisation of manganese may explain why manganese deficiency can occur on cereals during the early part of the spring when demand is high, but not later following solubilisation. Considerable variation in the soluble manganese content in the rhizosphere has been found, and this factor together with the problem of obtaining samples, particularly under dry soil conditions, means that it may be difficult to develop a diagnostic sampling procedure based on rhizosphere soil.

6.   Mobybdenum

The most striking feature of the molybdenum story is that crops can and do respond to molybdenum applications despite the small amounts required.

ARNON and STOUT (1939) who first demonstrated that molybdenum was an essential micronutrient, had great difficulty in purifying the growing media and nutrient salts before they could develop molybdenum deficiency symptoms on tomato, and were of the opinion that the essentiality of molybdenum was destined to remain a laboratory curiosity. But it was not long before ANDERSON (1942) in South Australia demonstrated spectacular yield increases on subterranean clover to ammonium bolybdate; the clue came from observations on the benefits of a molybdenum-rich ash from eucalypts.

Table 5.   Crops susceptible to molybdenum deficiency
Crops which are often treated are underlined

Alfalfa	Cauliflower	Peas
Beans	Clovers	Soybean
Broccoli	Lettuce	Spinach

6.1 Grass/Clover Pastures

Farmers in New Zealand were also lucky, in that the early research workers in the 1940s examined the need for molybdenum by clover which, in New Zealand, is the cornerstone of efficient herbage production; the result is the widespread application of 50 g/ha sodium molybdate every 4-5 years (as molybdenum enriched superphosphate).

Recent work in Australia, prompted by the finding of excessive storage of copper in sheeps' liver associated with low pasture molybdenum levels, has shown that despite the reduction to about 20 per cent, in efficacy of soil applied sodium molybdate (at 100 g/ha) after 2-3 years, there was sufficient in the soil to maintain a satisfactory plant level (BARROW et al. 1985).

6.2 Copper/Sulphur/Molybdenum

There is now general agreement that high levels of molybdenum, often in association with high levels of sulphur, inhibit copper utilization and retention by ruminants (e.g. SIMPSON et al. 1981, SUTTLE 1975). There is, in contrast, no general agreement on the mechanism involved and also of the way that iron plays a contributory part.

There are two extreme conditions that need to be considered. Firstly there is the situation in which the deficiency is due to low levels of copper in the herbage and not to the inhibition of copper absorption and utilization.

284

Secondly there are pastures such as the classical "teart" pastures of Somerset which contain very high levels of molybdenum, and where the symptoms are not those of typical copper deficiency. This latter condition should be thought of as a copper alleviated molybdenum deficiency and not as a molybdenum induced copper deficiency. The difficulty now is of deciding where copper deficiency ends and molybdenum toxicity starts.

A consideration of the molybdenum levels in pastures and fodder is frequently the key factor in understanding the reasons for variation in recommended minimum copper levels in herbage in different countries. Where, as in many parts of Australia and France, herbage molybdenum levels are somewhat low, herbage copper levels of around 7 ppm Cu are judged to be suitable; in contrast, in many parts of the UK where high molybenum contents prevail, higher herbage copper levels of at least 10 ppm Cu, are required.

A further interesting ramification of the molybdenum/copper story has been revealed in Norway, where there has been a recent increase in the incidence of chronic copper poisoning in sheep. It is postulated that the reduction in the use of lime and the associated decrease in soil pH has resulted in lower molybdenum levels in the herbage which in turn have permitted greater retention of copper in the sheep.

6.3 Molybdenum, Tungsten and Cancer

The inhibitory effect of dietary molybdenum on oesophageal and mammary cancer of rats, and the stimulation of cancer by tungsten has focused attention on the antagonism between tungsten and molybdenum; due to their similar size, tungsten may possibly replace molybdenum in some enzymes and impairment of enzyme activity by tungsten has been demonstrated in animal studies (SHIANG P. YANG 1985). The Chinese work on rats is particularly relevant in the tungsten mining areas in Jiangxi Province where breast cancer mortality is abnormally high.

6.4 Future

The molybdenum requirement for efficient nitrogen fixation in leguminous plants suggests that much more attention should be given to studies of the molybdenum nutrition on such crops as groundnuts, soybean, cowpea and the grams. Work in China, reported by LIU ZHENG (1984), has shown the way; substantial yield increases have been obtained by either seed treatment or foliar sprys with molybdenum on groundnuts, soybean, milk vetch, peas and lucerne as well as leguminous green manure crops.

7. Selenium

Selenium is unique amongst micronutrients in that in Finland selenium is now being applied in fertilizers with the express purpose of raising selenium intake by humans.

Selenium is also perhaps one of the most fascinating micronutrients. On the one hand it does not appear to be required by plants, although it has the potential to exchange with sulphur. On the other hand with livestock it has been the toxic effects that attracted attention until the essentiality of selenium was proved in 1957 (SCHWARZ et al.). Within the last 20 years it has been accepted, in many countries, that peroxidative damage of muscle fibres which results in a nutritional degenerative myopathy (white muscle disease) in livestock is caused by selenium deficiency.

Interest in selenium was given a considerable boost following the successful treatment with selenium in the Keshan region of China over the period 1974-1977 of a heart muscle disease which was common in young males, and could be fatal (XIAOSHU CHEN 1980).

Dietary intakes of selenium in China are the lowest recorded in the world (10-15 micrograms per day) and are followed by those in New Zealand (28-32 micrograms) and Finland (30-60 micrograms).

Studies in Finland have associated low selenium intakes with an increased risk of cardiovascular disease, and also with certain kinds of cancer. Although these studies are far from conclusive, the Finns decided in 1984 to enrich all fertilizers with selenium with the sole purpose of increasing the selenium level in the human diet, and hopefully thereby reducing the incidence of cardiovascular disease and cancer. It should be noted that there are only a few workers outside Finland who approve of this action, which should no doubt be taken as a pointer to attitudes in the event of fertilizer enrichment with say, copper or zinc being proposed as a means of increasing the amounts of copper and zinc passed down the food chain to man.

8. Zinc

There can be little doubt that the widespread occurrence of zinc deficiency on rice and the susceptibility of other important food crops particularly maize, to zinc deficiency means that it is the most serious micronutrient deficiency limiting world food production. In tropical America it is reported that zinc deficiency could limit production on over 700 million hectares (SANCHEZ et al. 1980). The cerrado soils of Brazil, which cover 180 million hectares, are of low zinc status. In South and South East Aisa it has been estimated that over 8 million hectares of irrigated rice suffer from zinc deficiency (PONNAMPERUMA 1982).

The susceptibility of beans and soybeans to zinc deficiency and the frequent occurrence of zinc problems on citrus and other fruit, further emphasizes the important position of zinc amongst micronutrients.

Table 6. <u>Crops susceptible to zinc deficiency</u>
Crops which are often treated are underlined

Apple	Flax	<u>Rice</u>
<u>Beans</u>	<u>Maize</u>	Sorghum
<u>Citrus</u>	Peach	Soybean
<u>Coffee</u>	Pear	Sudan Grass
Field Bean	Pecan	Tung

8.1 Rice and Breeding

In view of the enormity of the zinc deficiency problem on rice, the International Rice Research Institute in the Philippines conducted over the period 1971-1981 an intensive screening of 13,000 rice varieties for zinc deficiency. As a result of this work, in which 7 per cent were found to be tolerant to zinc deficiency, four new variaties, IR36, IR46, IR52 and IR54, were developed and released, which combine efficiency in absorbing zinc with high yield potential, as well as with disease and insect resistance. Yield advantages of up to 2 tons/ha on unamended soils can be obtained (PONNAMPERUMA 1982), and it is estimatewd that IR36 has already contributed an extra 5 million tons of rice grain each year since it was introduced.

Despite the improved zinc efficiency, the new varieties still respond to zinc treatment. The nature of the mechanism that confers improved zinc utilization is not known, but there are signs from work on cotton, groundnuts and sorghum in India that it will be possible to breed and select for zinc efficiency in these crops also.

8.2 Zinc and Human Nutrition

Zinc plays an essential part in over 100 enzymes of all categories in animals and, being a key element in DNA and RNA metabolism, is extremely important in all aspects of tissue development (VALLEE et al. 1981).

Whilst acute zinc deficiency in humans is limited to the classic cases in Egypt and Iran on males in late puberty, and to the condition of Acrodermatitis enteropathica caused by a genetic defect which impairs zinc absorption, there is considerable interest in the involvement of zinc in several aspects of human health. Zinc requirements are high during pregnancy and lactation, and some foetal malformations have been related to low maternal zinc status. Zinc levels in many western diets are low/marginal and vegetarians have a particularly low intake.

Apart from the various effects of zinc on sexual development in males, on skin, on foetal development, it is now postulated that zinc deficiency may be involved in anorexia. Zinc deficient anorexic monkeys lose weight but maintain normal plasma zinc levels as a result of the release of zinc from muscle tissue in an apparent attempt by the body to maintain a satisfactory zinc status despite a deficient intake.

8.3 Future

It is unlikely that zinc will lose its pre-eminent position as the most important micronutrient for food production worldwide, and all indications are that more attention will have to be paid to the maintenance of satisfactory zinc intakes in humans.

9. References

ANDERSON, A.J. 1942. Molybdenum deficiency in a South Australian ironstone soil. J. Aust. Inst. Agric. Sci. 8, 73-75.

BARROW, N.J., LEAHY, P.J., SOUTHEY, I.N., PURSER, D.B. 1985. Initial and residual effectiveness of molybdate fertiliser in two areas of South Western Australia. Aus. J. Agric. Res. 36, 579-587.

BIENFAIT, H.F. 1985. Regulated redox processes at the plasmalemma of plant root cells and their functions in iron uptake. J. Bioenerg. Biomembr. 17, 73-83.

BLAMEY, F.P.C. 1976. Boron nutrition of sunflowers (Helianthus annuus L.) on an Avalon medium sandy loam soil. Agrochemophysica 8, 5-10.

BOLLE-JONES, E.W., HILTON, R.N. 1957. Zinc deficiency of Hevea brasiliensis as a predisposing factor to Oidium infection. Nature, Lond. 17, 619.

BORCHMANN, W., GERATH, H. 1972. Effect of the micronutrient boron in potato growing. Arch. Acker-und Pflanz. und Boden. 16, 369-379.

BRANDENBURG, E. 1931. Die Herz-und Trockenfäule der Ruben als Bormangelerscheinung. Phytopath. 3, 499-517.

BRUCHLOS, P., BERGMAN, W. 1979. A contribution to the effectiveness of fertilization with micronutrients in the German Democratic Republic. Arch. Fur. Acker-und Pflanz. und Boden. 23, 1, 39-48.

DA SILVA, A.R. et al. 1983. Influence of micronutrients on the male sterility on upland wheat and on rice and soybean yield in red-yellow latosol. Pesq. Agropec. Bras. Brasilia 18, 593-601.

ELAD, Y., BAKER, R. 1985. Influence of trace amounts of cations and siderophore-producing pseudomonads on chlamydospore germination of Fusarium oxysporum. Phytopathology, 75, 1047-52.

FEHR, W.R. 1982. Control of iron deficiency chlorosis in soybeans by plant breeding. J. Pl. Nutrit. 5, 927.

FLOYD, B.F. 1913. Fla. Sta. Rep. 1912/1913. p.27.

GARTEL, W. 1974. Micronutrients - their significance in vine nutrition with special regard to boron deficiency and toxicity. Weinberg und Keller 21, 435-508.

GETTIER, S.W., MARTENS, D.C., BRUMBACK, T.B. 1985. Timing of foliar manganese application for correction of manganese deficiency in soybean. Agron. J. 77, 627-630.

GRAHAM, R.D., ROVIRA, A.D. 1984. A role for manganese in the resistance of wheat plants to take-all. Plant and Soil 78, 441-444.

HANNAM, R.J., DAVIES, W.J., GRAHAM, R.D., RIGGS, J.L. 1984. The effect of soil and foliar applied manganese in preventing the onset of manganese deficiency in Lupinus angustifolius. Aus. J. Agric. Res. 35, 529-538.

HANNAM, R.J., RIGGS, J.L. 1985. Fert Res. 6, 149-156.

HUDIG, J. et al. 1926. On the so-called reclamation disease as a third soil disease. Z. Pflanzen. Dung. 8A, 14.

KANNO, C. 1957. Studies on shrinking plant and failure in fruiting in rape plants. Bull. Div. of Plant Breed. and Cultiv. Tokai-Kiuki Nat. Agric. Expl. Stat. No. 5.

LEVESQUE, M.P., MATHUR, S.P. 1983. The efects of using copper for mitigating histosol subsidence on: 1. The yield and nutrition of oats and lettuce grown on histosols, mineral sublayers and their mixtures. Soil Sci. 135, 88-100.

LINEHAN, D.J., SINCLAIR, A.H. 1985. Mobilization of copper manganese and zinc in the soil solution of barley rhizospheres. Plant and Soil 86, 147-149.

LIU ZHENG, 1984. China: Status and Use of micronutrients. Micronutrient News, 4, 3, 1. Micronutrient Bureau.

MAFF, 1981. Survey of copper and zinc in food. Food surveillance Report No. 5, H.M.S.O. London.

MARTENS, D.C., PAYNE, G.G., PERERA, N.F. 1986. Crop responses to high levels of copper application. Int. Copper Res. Assoc. Ann. REpt. No. 292 (G).

MASCAGNI, H.J., COX, F.R. 1985. Evaluation of inorganic and organic manganese fertiliser sources. Soil Sci. Soc. Amer. Proc. 49, 458-461.

MATHUR, S.P., HAMILTON, H.A., LEVESQUE, M.P. 1979. the mitigating effect of residual fertilisation copper on the decomposition of an organic soil in situ. Soil Sci. Soc. Am. J. 43, 200-203.

MCDONNEL, J.G., FARRELL, E.P. 1986. The use of copper and small mineral soil additions in the control f the decomposition of peat. Private Comm.

MCGRATH, D., FLEMING, G.A., MCCORMACK, R.F. 1982. Effects of applying copper-rich pig slurry to grassland 2.land spreading trial. J. Agric. Res. 21, 49-60.

MORARD, P. 1985. Le cuivre et le ble. p63-86 in Proc. Int. Sympos. on Trace Elements in Agric. Foundation C.T.L. Ghent, Belgium.

NEAL, W.M. et. al. 1931. A natural copper deficiency in cattle rations. Science 74, 418-419.

OLSEN, R.A., CLARK, R.B., BENNETT, J.H. 1981. The enhancement of soil fertility by plant roots. American Scientist 69, 378-384.

PIPER, C.S. 1938. The occurrence of "reclamation disease" in cereals in South Australia. Austr. Council. Sci. Ind. Res. Pamphlet 78, 24-28.

PONNAMPERUMA, G.N. 1982. Genotypic adaptability as a substitute for amendments on toxic and nutrient deficient soils. Proc. of the Ninth Int. Plant Nut. Colloquium, Warwick, England (August 1982) p. 467.

REISER, S. et alia. 1985. Indices of copper status in human consuming a typical American diet containing either fructose or starch. Amer. J. Clin. Nutrit. 42, 242-251.

ROMHELD, V., MULLER, G., MARSCHNER, H. 1984. Localization and capacity of proton pumps in roots of intact sunflower plants. Plant Phys. 76, 603-606.

SAMUEL, G., PIPER, C.S. 1928. Grey speck (Manganese deficiency) disease of oats. J. Dept. Agric. S. Austr. 31, 696-705.

SANCHEZ, P.A., COCHRANE, T.T. 1980. Soil constraints in relation to major farming systems in tropical America. In: soil related constraints to food production in the tropics. I.R.R.I. Philippines p.109-139.

SCHWARZ, K., FOLTZ, C.M. 1957. Selenium as an integral part of Factor 3 against dietary necrotic liver degeneration. J. Amer. Chem. Soc. 79, 3292.

SHIANG P. YANG, HUI-JUAN WEI. 1985. Effect of molybdenum and tungsten on carcinogenesis, in Proc. 5th Int. Symposium on Trace Elements in Man and Animals (1984). C.A.B.

SIMPSON, A.M., MILLS, C.F., MCDONALD, J. 1981. Influence of dietary molybdenum and sulphur upon the utilization of copper by cattle. (Rowett Res. Inst.) Project 276, INCRA, New York.

SOMNER, A.L. 1931. Copper as an essential for plant growth. Pl. Physiol. 6, 339-345.

SPOONER, A.E. and HUNEYCUTT, H.J. 1983. Effect of boron ob coatal Bermuda grass. Arkansa Farm Res. 43,4,2.

SUTTLE, N.F. 1975. J. Agric. Sci. 84, 255-261.

TAHTINEN, H. 1970. Residual effect of boron fertilisation. Ann. Agric. Fenn 9, 331-335.

TEUTEBERG, W. 1978. Boron for rape. Proc. 5th Int. Rapeseed Conf. 1, 260-265.

TILLS, A.R. and ALLOWAY, B.J. 1981 Sub-clinical copper deficiency on crops in the Breckland of East Anglia. J. Agric. Sci. Camb. 97, 473-476.

VALLEE, B.L., FALCHUK, K.H. 1981. Zinc and gene expression. Phil. Trans. R. Soc. Lond. 294, 185-197.

VOSE, P.B. 1982. Iron nutrition in plants; a world overview J. Pl. Nutrit. 5, 233.

XIAOSHU CHEN. 1980. Studies on the relation of selenium and Keshan disease. Biol. Trace El. Res. 2, 91-107.

BORON IN THE PANNONIAN CHERNOZEM
OF THE VOJVODINA PROVINCE AND
ITS EFFECT ON SUGARBEET YIELDS

Prof. Dr. S. Manojlovic, Faculty of
Agriculture, Institute of Field and
Vegetable Crops, Novi-Sad

Sugarbeet is an important crop in Yugoslavia. Although it is
ranked only 13th in Europe according to its acreage of 121,400 ha
(statistical data for the period 1978-1981), Yugoslavia is actually
ranked fifth according to the yield of beets of 41.9 t/ha and to the
yield of biological sugar of 6.8 t/ha. In the period 1975 - 1985,
the Yugoslav sugarbeet acreage ranged between 107,000 and
150,000 ha, total production between 4.2 million to 6.8 tons
of beets, and yields ranged between 40 and 47 t/ha (Table 1). The
principal sugarbeet growing regions of Yugoslavia are Vojvodina
Province (S.R. Serbia) with an acreage of 84,000 ha, S.R. Croatia
(mostly the region of Slavonia) with about 30,000 ha, and Serbia
proper with 16,000 ha; in the other republics, sugarbeet is
altogether grown on 12,000 ha (data from 1983).

The most successful sugarbeet growers are in the Vojvodina
Province and in Slovonia where, in some localities, beet yields
reach up to 70 t/ha and yields of white sugar up to 12 t/ha. The
record yield of sugarbeet roots is 190 t/ha.

Sugarbeets are grown on the best soils. In the Vojvodina
Province, these are chernozem soils and chernozem-like soils; the
most important soil type in the latter group is meadow black soil
(Semigley). The soils in both groups are deep, usually formed on
loess or loess-like subtrates. The humic horizon in the chernozem
soils is 40 to 70 cm deep, with the transitive AC horizon reaching
from 25 to 40 cm. The content of humus in the surface layer of the
chernozem soils ranges usually between 3.0 and 3.5% and the total
nitrogen between 0.17 and 0.25%. These soils are usually calcareous
on the surface ($CaCO_3$ content in the layer of 0 to 20 cm
about 1 - 10%); in the transitive AC horizon, the content of
$CaCO_3$ reaches over 20%. The chernozem soils have a slightly
alkaline reaction: their pH in water ranges from 7.5 to 8.3, in KCl
from 7.2 to 7.5.

Unlike the chernozems which receive moisture exclusively from rainfall, black meadow soils (semigleys) have an additional moisture inflow at the bottom of the profiles which results in the formation of the gleisoil horizon at a depth between 150 and 200 cm. According to their morphological, chemical, and physical properties, the upper parts of the profiles are quite similar to the typical chernozem soil. However, their production properties are better than those of the chernozem soils, especially in dry years, owing to the additional moisture which reaches the root zone by means of capillary ascendence.

There are important complex questions regarding the production of sugarbeet on these soils and the further promotion of this production. Besides the question of application of basic nutrients (macroelements), there is the question of application of microelements, especially boron. It is well known that boron plays an important role in the nurition pattern of sugarbeet, pointed out in many research and technical papers (N. Petrovic et al., 1984; Milosevic et al., 1984).

A large number of researchers studied the effect of boron on sugarbeet yields and technological quality (Kastori and curic, 1970; Emelyanova et al., 1971; Ignatova et al., 1977, Rutskaya et al., 1978; Krunic et al., 1980; Rutskaya et al., 1980; Popovic et al., 1983; etc.). The explanation for this keen scientific interest is that sugarbeet belongs to the group of plants which are particularly sensitive to boron deficiency (Kastori, 1983). It is assumed that boron, by forming complex links with sugars, facilitates the transportation of sugars through cell membranes (Turnowska-Stark, 1960), thus enhancing the accumulation of sugar in the sugarbeet root.

The mobility of boron in the soil is considerably reduced by high contents of aluminium oxide, iron, and mineral clay, high pH, and by drought. It may lead to the occurrence of boron deficiency in plants. Soils in certain regions of Yugoslavia were found to be dificient in available boron (Kerin, 1963 and 1967; Ljubic, 1974). In these regions, sugarbeet responded well to the application of boron by increased root yields.

The calcareous chernozem in the Vojvodina Province is normally provided with boron. Low values were detected for leached and degraded chernozems (Kosanovic and Ruza Halasi, 1962; Radmila Kovacevic-Tatic, 1974). In addition to that, the current practice in the Vojvodina Province and in other parts of the country is to add boron-enriched mineral fertilizers to sugarbeets because, as the theoretical explanation states, sugarbeet yields could be lowered due to boron deficiency in intensively used soil. The objective of this paper is to confirm the above explanation and to examine the effect of boron application on root yields of sugarbeets grown on calcareous chernozem, the most frequent soil type used for sugarbeet growing in the Vojvodina Province.

In recent years, many authors emphasized the importance of micronutrients in intensive plant production (zeravica and Rajkovic, 1974; Stojanovic et al., 1978, 1979, 1980; Bergmann 1980; Ljubic, 1980; Kurbel, 1981). Opinions are divided regarding the effect of micronutrients on the yield and quality of sugarbeet grown on chernozem soils which are well provided with micronutrients. Results of a three-year study conducted by Kastori and curic (1970) and results of zeravica and Rajkovic (1972) indicate that the boron level did not affect the yield of sugarbeet grown on chernozem soil. Similar results were obtained in the region of Baranya (Todorovic et al., 1977). The latter study indicated that boron application actually tended to reduce the yield of roots and tops while the content of sugar was slightly increased. Conversely, the results of Krunic et al. (1980) showed that boron, manganese, copper, zinc, molybdenum, and cobalt affected positively the yield and quality of sugarbeets grown on chernozem and alluvial soils which were well provided with these elements. The largest effect was exerted by a combination of boron + manganese.

Three-year results of Ljubic (1980) indicate that boron, manganese, and zinc affect significantly the yield and quality of sugarbeets grown on pseudogley and loessivized brown soil. According to Kurbelova (1981), the highest yields of roots for the region of Slavonia are achieved after the application of 6.07 - 6.58 kg/ha of boron and highest sugar yields with somewhat higher doses, 6.32 - 6.85 kg/ha of boron. It was concluded on the basis of more than 500 trials and production experiments conducted recently in the German Democratic Republic (Bergmann, 1980), on soils poor in micronutrients, that the application of micronutrients brings high yield increases. According to this author, it is necessary to pay attention to the content of available micronutrients in different soils in order to intensify plant production.

The extension service of the "Backa" Sugar Refinery at Titov Vrbas, Yugoslavia, pays great attention to sugarbeet nutrition with boron on the production area of the factory. The production area located in central Backa, one of three sub-regions of the Vojvodina Province, covers about 140,000 ha of arable land. Sugarbeet is grown on about 9,000 ha. About 50% of the production area belongs to the social sector. The average sugarbeet acreage in the social sector is over 10%, in the private sector about 3%.

Based on data from literature and on the results of the technical services of the fertilizer industry, the recommendation for boron fertilization included the foliar application of 2 - 4kg B/ha or boron in combination with compound fertilizers.

The importance of solving this problem is emphasized by the fact that fluctuations and even reductions in yields were experienced in the last decade. Graph 1 shows the large variability in the yields of beets and sugar with a large drop in 1982, due to Cercospora attack, and in 1983, due to an excessive drought. Graph 2 shows three-year average yields of beets and sugar obtained in the production area of the "Backa" Sugar Refinery.

Since the extension service and laboratory of the factory have recently joined the System of Soil Fertility and Fertilizer Application Control of the Vojvodina Province, it would be of interest to consider critically their results and make use of them.

Our own methodological approach to the solution of the problem of the role of boron in sugarbeet production on Pannonian chernozem consists of the following:

a) following the yields of beets and sugar in the production area by estimating yield and technological quality for over 30 one hectare plots distributed evenly over the entire production area. Yields are estimated three times in the course of the growing season: in June, around August 15, and in late September;

b) following the elements of soil fertility by assessing the levels of available nitrogen (nitrate and ammonium), phosphorus, potassium, natrium, and exchangeable calcium and magnesium, the levels of available micronutrients: boron, manganese, zinc, copper, cobalt, and iron, as well as the levels of heavy metals, cadmium, lead, and pickel. These analyses are conducted four times in the course of the growing season: in March, before planting and emergence, before the beginning of the intensive uptake in late May, before the beginning of intensive sugar accumulation in mid-September, and at the end of the season, in late September.

In addition to that, pH value, humic content, and total nitrogen content are determined once a year, i.e. in late May.

c) based on analytical data, relationships are calculated between the content of available boron on the one hand, extracted by boiling water as a dependent variable, and pH in water and M KCl, humic content, total nitrogen content, and the contents of available phosphorus, potassium, natrium, exchangeable calcium, and magnesium as independent variables, on the other hand. Correlations are determined on the basis of the average value for these parameters from four assessments per growing season. The correlations are determined for the plowing layer and for the entire soil profile. Furthermore, mineral nitrogen, as an independent variable for each date of sampling, is correlated with the average boron content in the soil.

d) relations are determined between boron contents in the soil layer of 0 to 30 cm and the entire profile as independent variables, on the one hand, and production results, as a dependent variable, on the other; the production results are expressed as the yield of roots, sugar content (%), and the yields of sugar and recovered sugar (Rdt) in tons/ha.

e) relationships are determined separately between the independent variables, i.e. available boron in the soil and nitrate nitrogen in the soil, and a dependent variable, i.e. amount of boron (g. h-1) taken up by the whole plant.

f) a special approach to the solution of the problem studied was worked out for establishing the effect of natrium tetraborate (Borax), foliar applied, on the yield and quality of sugarbeet.

To that effect, two identical trials are established each year, one in irrigation and another one in dry farming conditions. Borax is applied at the amount of 2 and 4 kg B/ha in a single spraying.

RESEARCH RESULTS

It has been mentioned before that all experiments dealt within this report were conducted on production plots. Each year we selected 20 one-hectare plots for each soil type - chernozem soil and black meadow soil (semigley). In each plot we estimated the yields of beets, sugar, and recovered sugar, and the percentage of digestion (the percentage of sugar in the root). In that way, each data represents the average of 20 assessments per soil type. Yield parameters were assessed three times in the course of the growing season: in June, mid-August, and late September (at the end of the season). In this report, we take into consideration only the yield at the end of the season.

However, for the sake of a broader insight into the yield performance of the entire production area, amounting to about 9,000 ha, we present data on the average yields of beets and sugar for the production area, regardless of the soil type, in Graph 1 (separately for each year). The same parameters are presented in the form of three-year average data in Graph 2.

Graphs 1 and 2 show that the yields of beets ranged from 40 to 52 t/ha; large reductions took place in 1982 (about 45 t/ha, due to Cercospora attack) and in 1983 (41 t/ha, due to excessive drought).

The contents of available boron in the soil layer of 0 to 30 cm and in the layer of 0 to 150 cm.

Table 2 shows the contents of available boron in chernozem soil and Table 3 the contents in black meadow soil (semingley). Boron contents in either layer varied largely in both soil types. However, there were no essential differences if the examined soil types were considered according to their average values.

The percentage of digestion was negligibly higher in chernozem soil than in black meadow (semigley) (Tables 2 and 3).

However, all yield parameters (yields of beets, sugar, and recovered sugar) were somewhat higher on black meadow soil (semigley) (Tables 2 and 3). This appears to be due to the additional moisture entering the lower part of the soil profile. All parameters varied largely with both soil types, from plot to plot as well as from year to year.

EFFECT OF AVAILABLE BORON IN THE SOIL ON YIELD PARAMETERS
(data obtained under production conditions)

It is important to establish a connection between the level of available boron in the soil and sugarbeet yield parameters in order to obtain the answer whether the PRESENT LEVEL of available boron is sufficient for high and stable production of sugarbeet or whether it is necessary to resort to boron fertilization.

It is well-known that great difficulties are encountered in determining the relationship between individual factors in natural conditions because of their large and uncontrollable variations.

Our experiments were conducted under natural conditions and it is thus reasonable to expect that the correlations between the factors observed will naturally be lower.

As regards the contents of boron in the soil, samples were
taken four times in the course of the growing season while the
results are given as average values of the four assessments and
compared with the yield at the end of the season obtained in 12
selected plots each with chernozem and black meadow soil (semigley).

RELATIONSHIP BETWEEN THE CONTENT OF AVAILABLE BORON IN THE SOIL AND THE PERCENTAGE OF DIGESTION (THE PERCENTAGE OF SUGAR IN THE BEET)

Table 4 shows that no correlation could be found between the
contents of available boron in the layers 0 - 30 and 0 - 150 cm for
the chernozem soil. For the semigley soil, however, such
correlation was found in 1985, as expressed by corresponding
indicators of linear, exponential, and logarithmic regression.

RELATIONSHIP BETWEEN THE CONTENT OF AVAILABLE BORON IN THE SOIL AND THE AMOUNTS OF SUGAR AND RECOVERED SUGAR

Tables 6 and 7 show that there existed correlations between the
contents of available boron in the layer of 0 to 30 cm (chernozem
soil) and the amounts of sugar and recovered sugar in two out of the
three research years (1984 and 1985).

With the same soil type, the content of boron in the layer
of 0 to 150 cm, on the one hand, and the amounts of sugar and
recovered sugar, on the other, were correlated in one research year
only (1984).

It may be concluded on the basis of the above data on the
relationships between the content of available boron in the soil (in
the layers 0 - 30 and 0 - 150 cm) on the one hand, and the estimated
yield parameters (the percentage of digestion, beet mass, sugar
mass, recovered sugar mass), on the other, that the above parameters
do develop certain correlations in some years. This means that the
level of available boron in the soil must be taken into account
when making decisions on the application of boron.

RELATIONSHIP BETWEEN THE COMPOSITION AND PROPERTIES OF SOIL AND THE CONTENT OF AVAILABLE BORON IN THE SOIL

Data from literature state that there is a considerable number
of soil factors which affect, to a varying degree, the content,
solubility, and availability of boron to plants. The effect of
individual factors tends to become more pronounced if largely
different soil materials, regarding their composition and
properties, are analysed.

Our study, however, included only two soil types which are quite similar, i.e. chernozem soil and black meadow soil (semigley), which differ only in the moisture pattern in the bottom part of the profile. This is why no significant differences could be found in respect to composition and properties between the two soil types. Under production conditions, semigley is slightly more favourable which results in somewhat higher yields than those obtained on chernozem soil.

THE USE OF BORON FOR FERTILIZATION RECOMMENDATIONS FOR FURTHER WORK

Many authors studied the effect of boron application on chernozem soils medium to well provided with boron (e.g. N. Petrovic et al., 1984; Krunic et al., 1980; R. Milosovic et al., 1984; etc.). They found that boron either does not at all or only slightly affect the beet yield. The field experiments carried out by the sugar refinery from Titov Vrbas confirmed these results.

The three-year results (Tables 2 and 3) showed for both soil types, chernozem and semigley, at the production area of the factory, that they are medium to well-provided with boron. This means that no significant effects of boron application should be expected and that no boron fertilization should be practised. To prove the above statement beyond doubt, however, it is necessary to undertake a more extensive research data of boron application under different agroecological conditions.

This would ensure that the System of Soil Fertility and Fertilizer Application Control, introduced to the Vojvodina Province as the most intensive agricultural part of the country, gains its full scientific and economic justification. As the system has recently been legally accepted as the scientific basis for a rational use of fertilizers, i.e., the control of factors of plant production, it may be expected that the system will contribute to a rational use of both macro- and micronutrients, including boron, in the production of one of the most intensive field crops - the sugarbeet.

REFERENCES

1. Kastori, R., Curic, R. A study of the effect of complex fertilizers with addition of microelements on the yield and quality of sugarbeet, Agronomy Herald, 9, 231 – 236, 1970.

2. Kastori, R. The role of elements in plant nutrition, Matica srpska, Novi Sad, 1983.

3. Kosanovic, V., Ruza Halasi. Boron in the soils of Vojvodina Province, Annals of Research Papers of the Faculty of Agriculture, Novi Sad, Vol. 6, 1 – 9, 1962.

4. Kurbel Olga. Economic analysis of the effect of boron fertilization on the yield and quality of sugarbeet, Agronomy Herald, No. 5 – 6, 525 – 544, 1981.

5. Ljubic, J. Problems of boron deficiency, Economy, 18, 27 – 29, 1974.

6. Miljkovic, N. The origin of boron in the salinas of Vojvodina province, 1960.

7. Milosevic, R., Stanacev, S., Ubavic, M., Petrovic, N. Effect of boron, copper, zinc, and manganese on the yield and quality of sugarbeet, Soil and Plant, Vol. 33, No. 3, Belgrade, 1984.

8. Petrovic, N., Kastori, R., Ubavic, M. Effect of boron on the yield of sugarbeet, Contemporary Agriculture, Vol. 32, No. 9 – 10, Novi Sad, 1984.

9. Todorcic, B., Crnogorac, S., Faller, N., Bertic, B. Dynamics of microelements B, Zn, Cu, and Mn in sugarbeet roots and leaves in the course of growing season, determined in the region of Baranya, Soil and Plant, Vol. 26, No. 1, 91 – 102, 1977.